Hard Math

for

Middle School:

IMLEM Plus Edition

Glenn Ellison

ISBN-10:   1-4538-1445-0
EAN-13:   978-1-4538-1445-1

To Caroline, Anna, Kate,
and the Bigelow Math Team

Contents

IMLEM Meet #4 119

IMLEM Meet #5 151

Introduction

This is a book for sixth graders who can't believe that they're being asked to do yet another worksheet full of problems like: "Find -2 + -3." It is a book for kids who have just spent three weeks learning to draw factor trees, but never seen anything even remotely interesting that you can do with a factor tree. It's a book for kids who can't believe that in the sixth month of their algebra class the teacher is still going over how to solve $2x - 3 = 5$. It's a book for kids who want to finally do something hard. To be asked questions like:

How many factors does 1,000,000 have?

What is the 258th digit after the decimal place when 1/17 is written as a decimal?

This roll of Scott Tissue toilet paper has 1000 sheets. Using an ordinary ruler find the thickness of a sheet of toilet paper to the nearest thousandth of an inch.

It's a book for kids who are jealous that there are some schools in which problems like this are taught. Kids who are jealous and that there are some kids in those schools who could figure out answers to the above problems in their heads in 20 seconds.

This is also a book for teachers who would love to do interesting things with the students in their classes for whom everything comes easily. Teachers who would love to do so, but who are given zero resources by their school and have limited time in their crowded school day to find material for advanced kids. It is for teachers who have sometimes given their seventh-graders an eighth grade textbook, but realize that there's little point to this: the questions in eighth grade book aren't hard either, and working on them now will just make a seventh-grader more bored next year in eighth grade.

If you glance quickly though this book you'll note that the organization is odd. I jump back and forth from geometry to arithmetic to number theory to algebra. A few sections repeat material that was just covered thirty or forty pages earlier. There is a method to this madness. You are reading the "IMLEM Plus" edition of *Hard Math for Middle School*.

The Intermediate Math League of Eastern Massachusetts (IMLEM) has inspired junior high and middle school students for over forty years. Five times a year, kids participating in IMLEM travel as a team to nearby schools where they compete against those schools (and remotely against other schools) in mathematical problem solving. Kids who get 97% or higher on every test they take in school struggle to break 60% on the IMLEM meets and love it. IMLEM is a curricular math league: each meet covers an announced set of topics. The meets build camaraderie and give a real sense of accomplishment. The topics are well thought out. Even if a teacher just teaches to the test, kids will learn a tremendous amount.

MathCounts® is a national math club and math competition program. It's about 30 years old and has about 6000 participating schools. Most schools in Massachusetts do just one of IMLEM or MathCounts. But I'm hoping that the release of IMLEM Plus will induce at least a few more schools to do both. Mathcounts is a lot of fun and not at all hard to add to your math team calendar. MathCounts harder to study for: they don't announce what topics will be covered and each contest covers a very broad range. But by reading through old tests you can get a pretty good idea of what topics kids should try to learn. There's enough overlap between IMLEM and Mathcounts so that IMLEM students who don't forget each topic right after the corresponding meet (and who know algebra well) should be pretty well prepared by the time MathCounts season arrives.

Although the book is organized around the IMLEM calendar, I think of it as a book that could be used in (or out of) any school where there's a desire to learn some harder math. I've thought about writing a third version of this book with the sections in a more mathematically dictated order, but don't know that it would be any better (and I also have a job). The cycling from topic to topic to topic has a nice rhythm to it, keeps things fresh, and makes for a nice year-long math enrichment course. The occasional repetitions solidify kids' understanding of the most important topics. The fact that IMLEM is so open and public-spirited also makes it essentially as if there's an online enrichment activity that comes along for free with every copy of this book. Even if you're not fortunate enough to be able to compete in IMLEM, you can download many old competitions from http://www.imlem.org. Each new contest is also posted there a few weeks after it occurs. One could treat these like online contests and compare one's scores with the posted score distribution from the IMLEM schools.

Mostly, of course, this is a book for those who are participating in IMLEM and MathCounts. There are other books that cover parts of the IMLEM material, but figuring out what exactly is covered on each meet would take much more time than most vastly underpaid math team coaches could possibly put in. This book is designed to make running an IMLEM + MathCounts math team as easy as possible. Each chapter is divided into four sections. All you need to do to have a good IMLEM team is to cover the four sections in the four weeks between IMLEM meets. The first five chapters correspond page-for-page with the older IMLEM edition of this book, so coaches can let kids can buy either edition and use them interchangeably. Of course, just reading this book won't make anyone into a great problem solver. To get good at solving problems you need to solve problems. The IMLEM website is a great resource for this.

The sixth chapter of this book is designed to help IMLEM math teams get ready for MathCounts. It covers some topics that come up in MathCounts but are not covered in IMLEM. A word of warning, however, is that the sixth chapter of this book is harder than the first five. You should not let this deter you from entering MathCounts. Mathcounts contests are nicely designed to be accessible to kids from a wide range of backgrounds. If an IMLEM team was pressed for time it could enter MathCounts and do pretty well without any extra preparation at all. But it's unlikely that anyone would win MathCounts without knowing the material in the sixth chapter. So I thought it would be good to include it.

This IMLEM sections of this book also include a fair amount of material that has never appeared on IMLEM meets. Some are things they might include in the future. Some are things related to IMLEM topics that seem to come up on MathCounts. Others are things you're unlikely to ever need to know for middle school math contests. I've included them because I think they're interesting and kids who like math would enjoy reading them. Some of the sections too long to cover in a math team meeting if you include everything, so I've put the words "Advanced Topic" in front of things that are more advanced and unlikely to be of immediate use in IMLEM.

I'd like to thank all the people who've worked so hard over the years to make IMLEM what it is. Special thanks go to my older daughters, Caroline and Anna, to my wife, Sara, to Dan Fudenberg, and to Kent Findell. Caroline and Anna helped out with proofreading, checking, and editing. Sara designed the cover with help from Carol Fisher. Dan proofread an earlier manuscript, made a number of useful suggestions, and made up many of the figures. Kent wrote most of the IMLEM questions that are used as examples in this book. They're a tremendous resource. I hope he appreciates that any criticisms I've included in the text are just there because I thought they'd appeal to twelve-year-olds. The IMLEM problem-writing job has changed hands a couple of times since I wrote the first version of this book. Josh Frost took over the question-writing job in 2007-2008 and did it for a couple years before passing the job on to Avi Nagar. I haven't examined Avi's contests closely enough to see if there differences in how he interprets the topics that might be useful to cover. Perhaps I will in some future year. The one thing I have noticed is that he seems to have made the team rounds harder. Personally, I think this is good and that they were both a good challenge for the kids and a way to keep the final outcome of a meet up in the air until the very end.

I'd also like to thank anyone who's making the effort to read this book. It won't take you long to realize that this book is not like your other middle school textbooks. Specifically, you'll notice that it's hard. Sometimes it'll take you fifteen minutes to do the problems you're supposed to do before going on to the next paragraph. Other times you'll decide you need go back to the very beginning of the section you're almost done with because you suddenly realize you have no idea what's going on. I appreciate that you're willing to make this effort and hope you enjoy it.

Finally, I'd like to thank the Bigelow coaches, my daughters, and all the kids who have read earlier versions of this. No one in their right mind would write a book like this to sell copies. There aren't enough kids in IMLEM for me to sell many copies. And I've set the price sufficiently low so that I make very little per copy. I wrote this book because I thought my daughters and their teammates would like to read it. It's been great to see the Bigelow team making use of it and doing so well. And it's been even more surprising and equally gratifying to see how many copies Amazon is somehow selling to kids with no connection to IMLEM. It's because I assume that many of these copies are being bought by kids who do MathCounts that I did get around to adding the extra chapter to expand the IMLEM edition to IMLEM Plus.

IMLEM Meet #1

Welcome to IMLEM! The IMLEM season consists of five meets for which you'll travel to nearby schools. For each meet an IMLEM team must choose ten members whose scores will count. Other students can come along as alternates. Going to a meet as an alternate can be even more fun than going as an official competitor: you get to go on the same bus ride; you can take every round just like an official competitor would; and you have less reason to be nervous. If your school doesn't encourage participation by alternates, tell them they should.

Each IMLEM meet consists of six rounds. Five are individual rounds. You get ten minutes to do three questions by yourself. Each student is "official" on three of the five rounds. The team must have same number of official participants in each individual round.

Question: How many team members are official in each individual round?

There is also a team round. The team works together to solve six (harder) questions in fifteen minutes.

There five individual rounds for each meet are: Mystery; Geometry; Number Theory; Arithmetic; and Algebra. Mystery is a mystery. You never know what will be on it. The other four rounds have much more specific announced topics, e.g. the number theory category in meet 1 covers prime numbers, factors, and divisibility rules. The IMLEM edition of *Hard Math* is arranged to make it as easy as possible to learn the material you'd need to know to do well on IMLEM. Each section covers what's on the corresponding IMLEM round.

Students who are new to IMLEM may want to just try to learn the material on two or three categories. This is only advisable, of course, if they're going to be official in the categories they studied. This requires either that your team divide up the categories well in advance, or that older students try to learn all of the categories so they can be more flexible.

Category 2 – Geometry

The geometry problems in meet #1 all seem to ask you for the measure of some angle in degrees. The hardest thing about the category is that there are a fair number of definitions and facts you need to know to do the problems, and it takes a bit of practice to get used to figuring out which facts to use. A nice feature of the category is that the problems can usually be done pretty quickly if you know all the facts and get the hang of using them.

2.1 Basic Definitions

- An *angle* is formed by two rays emanating from a single point. The point where the rays come together is called the *vertex* of the angle. In the picture below B is the vertex of the angle. The angle below would usually be called angle ABC. Sometimes people shorten this to "angle B", but I think this is a bad idea because things can get confusing when different angles share a vertex.

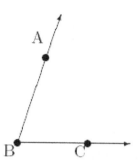

- The standard way to measure angles is in *degrees*. The figures below give some examples.

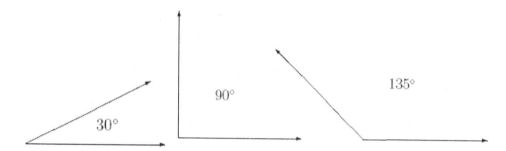

- An angle that measures less than 90 degrees is called an *acute* angle.

- An angle that measures 90 degrees is called a *right* angle. The angles in a square are all right angles. A standard way to show that an angle in a picture is a right angle is to draw a little box where the lines meet.

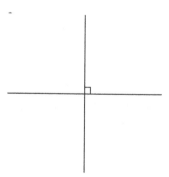

- An angle that measures more than 90 degrees is called an *obtuse* angle.

- You can think of a straight line as a 180 degree angle.

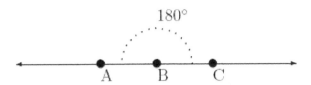

2.2 Adding Up Rules

Most IMLEM problems in this category tell you the measure of one or two angles in a figure and ask you to figure out the measure of some other angle.

The way to get the answer in some problems is to figure out the unknown angle using adding-up rules. There are three main adding up rules you need to know.

- The sum of all the angles around a point is 360 degree. For example, in the figure below, the adding-up-to-360 rule says that the measures of angles GRY, YRT, TRM and MRG add up to 360 degrees

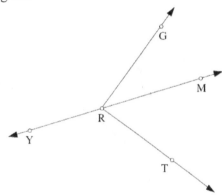

A common way to make a math contest problem out of a rule like this is to tell you the measures of some angles and ask you to find another. For example,

Question: In the figure below CAD is 80 degrees, DAE is 100 degrees, and EAB is 80 degrees. Find the measure of angle BAC.

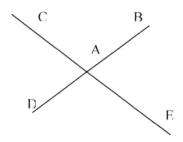

Answer: 100 degrees. The sum of the measures of the other three angles is 80 + 100 + 80 =260. The sum of the measure of all four angles is 360 degrees, so BAC must be 100 degrees.

- Two angles that together make up one side of a straight line are called *supplementary* angles or *supplements*. For example, in the figure above DAC and CAB are supplements. The measures of supplementary angles add up to 180 degrees.

  Question: In the figure above CAD is 80 degrees. Find the measure of angle BAC.

  Answer: 100 degrees. CAD and BAC are supplements. 80 + 100 = 180.

  Question: In the figure below what is the measure of angle CAD if it is less than 180 degrees.

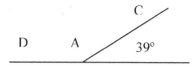

Answer: 141 degrees. Angle CAD and the unnamed 39-degree angle are supplements. 39 + 141 = 180.

You may be wondering why the question bothers to tell you that the measure of CAD is less than 180 degrees. The reason is that there would technically be another possible answer if they didn't. The points C, A, and D actually make up two different angles. One above and to the left of A that measures 141 degrees and another below and to the right of A which you can think of as going the long way around from C to D that measures 219 degrees. Usually, when people ask about an angle formed by three points they mean the smaller one even if they don't say it, but if you're trying to do a problem and something is going wrong you might pause to think whether this could be a problem where they're actually asking about the larger angle.

- Two angles that add up to a right angle are called *complementary* angles. Complementary angles add up to 90 degrees. If the angle that looks like a right angle in the figure below really is a right angle and a and b are the measures of the two angles in degrees then a + b = 90.

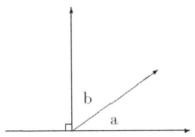

2.3 Equality Rules

In some problems just knowing the three adding-up rules is enough. In other problems you also need to know some equality rules. For all of the meets they've had in this millennium, you would have done fine if you just knew two of these.

- Opposite angles are equal.

    Opposite angles are angles that have the same vertex AND the same sides. For example in the figure I used to illustrate adding up to 360°, CAD and BAE are opposite angles. That's why they were both 80°.

- Corresponding angles on parallel lines are equal.

    Two lines in a plane are *parallel* if they would never meet no matter how long you made them. Corresponding angles are angles formed when a single line intersects two parallel lines in the same way.

    In the picture below, for example, the two lines that slope down are parallel. The angles marked with the double arcs are corresponding angles.

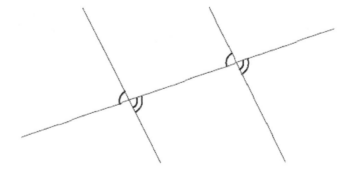

    Here's another example. In the figure below, if line m and line n are parallel then angles HKR and QRS are corresponding angles.

If you find it a little hard to identify corresponding angles, that's normal. I think this rule is harder to apply than any of the others I've discussed.

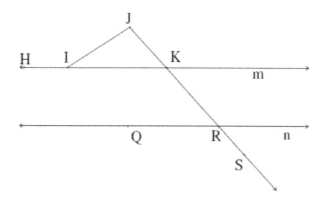

Question: (IMLEM Meet #1, Oct. 2003) Lines m and n are parallel. Angle HIJ is 148 degrees. Angle QRS is 133 degrees. Find the measure of angle IJK.

Answer: 101 degrees. We get this by applying several of our rules in succession. First, IKR and QRS are corresponding angles on parallel lines, so the measure of IKR is also 133°. Next, we use the supplements rule twice: IKR and IKJ are supplements so IKJ = 47° (47 + 133=180); and HIJ and JIK are supplements so JIK=32° (148 + 32=180). Finally we use a rule that I haven't told you about yet: the angles in a triangle add up to 180°. In this problem, we know that the measures of the two angles at the bottom of triangle IJK are 32° and 47°, so it must be that the angle at the top is 101°. (32 + 47 + 101 = 180).

A related fact that's needed for some problems is:

- Corresponding angles on two lines are equal if and only if the lines are parallel.

For example, if they'd wanted to make the question above a little harder they could have asked:

Question: Angle HIJ is 148 degrees. Angles QRS and HKR are both 133 degrees. Find the measure of angle IJK.

The first step in doing this problem is to say if angles QRS and HKR are equal, then lines m and n must be parallel. From there it's the same problem. The answer is therefore still 101 degrees.

2.4 Angles in Polygons

One rule for angles in polygons you know is useful because I've just used it:

- The angles in a triangle add up to 180°.

The generalization of this formula beyond triangles turns out to be equally important.

- A four-sided figure is called a quadrilateral. Squares, rectangles, parallelograms, rhombuses and trapezoids are all quadrilaterals. The angles in any quadrilateral add up to 360°.

- A more general formula that's very good to memorize is that the angles in an *n*-gon add up to (*n*-2) 180°. For example, the angles in any 12-gon add up to 1800°.

A polygon is said to be *regular* if all sides are of equal length and all angles are of equal measure.

- A corollary of the above fact is is that the measure of each angle in a regular *n*-gon is $\frac{n-2}{n}$ 180°. For example, each angle in a regular octagon is (6/8) × 180 = 135°.

If you forget the *n*-gon angle formula there are a couple ways to figure it out. One way (as illustrated in the picture on the left below) is to put a point directly in the middle of an n-gon and draw lines connecting it to each of the vertices. In a regular pentagon, for example, there will be five angles around the point in the middle. The five equal angles in the middle add up to 360°, so the measure of each one must be 360°/5=72°. The angles in each of the five triangles we've made must add up to 180. Hence, the two angles on the outside of each triangle must be 54°. (72 + 54 + 54=180.). Each angle of the pentagon is a sum of two of these angles, so each angle in a regular pentagon is 108°. You can then calculate that the sum of the five angles in the pentagon is 5 ×108 = 540.

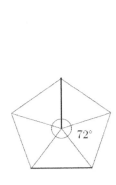

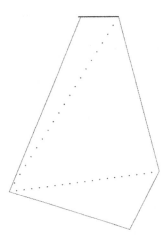

A second way to get this answer that works for nonregular pentagons too is to divide the pentagon into triangles by drawing lines connecting one vertex to each nonadjacent vertex. The picture on the right illustrates that doing this divides a pentagon into three triangles. The angles in each triangle add up to 180°. Each angle in the pentagon is a sum of one, two, or three angles from these triangles. Each angle in the triangles belongs to exactly one pentagon angle. Hence, the sum of the angles of the pentagon is exactly equal to the sum of the angles in all three triangles. 180 ×3 = 540.

While these are neat arguments, neither is something you'll want to do in the middle of a math meet in which you only get 3 minutes to answer each question. The way to avoid this, of course is to MEMORIZE THE N-GON ANGLE FORMULA!

- If you're having a hard time memorizing the formula here's an easier version: the n exterior angles of an n-gon add up to 360°. The only thing that's difficult about this one is to remember what an *exterior angle* is. I think of it as the angle that a polygon makes when you put it on top of a table. The diagram below shows an example:

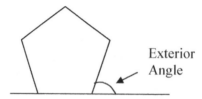

- In a regular n-gon, the fact that exterior angles add up to 360° implies that the measure of each exterior angle must be $360/n$. For example, in a hexagon (6-gon) this tells us that each exterior angle is 60°. Interior and exterior angles are supplements (remember what that means?), so the interior angle is 120°. In a dodecagon (12-gon), each of the exterior angles is 30°. The interior angles are 150°.

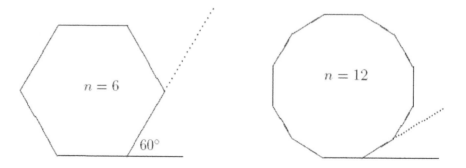

Some IMLEM questions are really trivial if you know this formula. For example,

Question: (IMLEM Meet #1, Oct. 2000) How many degrees are in an exterior angle of a regular 18-gon.

Answer: 20°. 360 / 18 = 20.

The table below lists the interior angles for many different *n* gons. If you're having trouble memorizing the *n*-gon formula, another strategy would be to just memorize the table below. Caroline thinks this would be harder than memorizing the formula. I think she's right. If you disagree though, you could do memorize the table. You could also decide to do it even if you have memorized the n-gon formula. During the meet it would let you save the time it takes to do the division.

n	Popular name for *n*-gon	Sum of interior angles	Interior angle in regular *n*-gon	Exterior angle in regular *n*-gon
--	-------	-------	-------	-------
3	Triangle	180	60	120
4	Quadrilateral	360	90	90
5	Pentagon	540	108	72
6	Hexagon	720	120	60
8	Octagon	1080	135	45
10	Decagon	1440	144	36
12	Dodecagon	1800	150	30
15	15-gon	2340	156	24
18	18-gon	2880	160	20
20	20-gon	3240	162	18
n	*n*-gon	$180\,(n-2)$	$(n-2)/n\,\,180$	$360/n$

2.5 Problem Solving Strategies

Only a few IMLEM problems are as simple as the one above. The one I did before that, which required you to apply three different facts in sequence, is more typical.

Here are some strategies you might use to go about solving the problems (and to try to solve them quickly enough so that you can work on more than just one problem in the ten-minute time limit).

> *2.5.1 Don't worry about figuring out how to get the answer.*
> *Just fill in whatever you can fill in until you get there.*

In many questions you'd drive yourself crazy if you tried to think about exactly how you'll solve the problem: will you learn what you want if you first use a triangle-adding up rule to get one angle, then do a supplementary angle, then do an opposite angle, then do a corresponding angle?; or will this not work so you should try to first fill in the opposite angle, then add up the angles in a quadrilateral, then do something else, etc.?

One reason why you shouldn't do this (in addition to the fact that we'd like you to keep your sanity), is that in many IMLEM angle problems there are lots and lots of ways to get the answer. You're likely to eventually get to the answer almost regardless of how you try to get there as long as you don't waste too much time planning before you start solving. For example, consider

> *Question: (IMLEM Meet #1, Oct. 2002) In the figure shown below angles CAE, GFE and CDB are right angles and angle ACE measures 27 degrees. How many degrees are in the measure of angle DGF?*

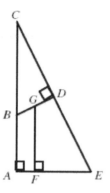

One way to get the answer would have been to first use the triangle adding up rule in triangle ACE: the fact that $27 + 90 + AEC = 180$ implies that $AEC = 63$. From here, one could use the quadrilateral adding up rule in quadrilateral DEFG: $90 + 63 + 90 + DGF = 360$. This implies that the measure of DGF is 117 degrees.

Another way that you might have started would have been to use the triangle adding up rule in triangle CBD: the fact that $27 + 90 + CBD = 180$ implies that $CBD = 63$. CBD and DBA are supplements, so you could next fill in that the measure of DBA is 117 degrees. At this point you could either (1) notice that DGF and DBA are corresponding angles on parallel lines so that DGF must also be 117 degrees or (2) use the quadrilateral rule in AFGB ($90 + 90 + 117 + BGF = 360$) to find fill in that the measure of BGF is 63, and then use the fact that BGF and DGF are supplements to find DGF. (Believe it or not, this also gives 117).

Some paths to the answer turn out to be a little longer than others. But, solving these problems is not like finding the solution to a maze. There aren't a lot of dead ends. Just go down some path and you're likely to get to the end.

Here's another example:

> *Question: (IMLEM Meet #1, Oct. 2003) Lines TP, BG, and DM intersect at point O. The measure of angle BOT is 47°. The measure of angle MOG is 29°. How many degrees are in the measure of angle DOP.*

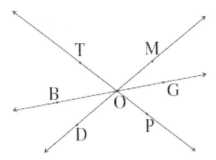

This problem can be done in lots of ways by using the opposite angles and supplements rules in different orders. Try a couple. They should all give 104° as the answer.

2.5.2 Find angles by subtraction

One useful thing to remember when you're trying to figure out angle measures is that you can sometimes fill in measures by subtraction. For example

Question (IMLEM Meet #1, Oct. 2001): In the figure below regular pentagon AGHIF sits inside regular hexagon ABCDEF so that the two shapes share base AF. How many degrees are in the measure of angle GAB?

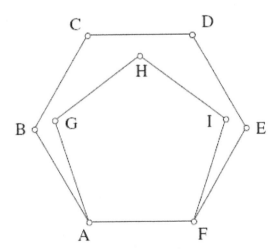

Answer: 12. Angle BAF is 120° because it's the interior angle in a regular hexagon. Angle GAF is 108° because it's the interior angle in a regular pentagon. You can get the measure of angle BAG by subtracting these, 120-108=12.

Here's a challenge problem using the same figure:

Question: In the figure above, what is the measure of angle GAE?

23

I know several ways to do this using facts I haven't discussed yet. One uses strategy *2.5.4 Draw in missing lines* and advanced fact 2.6.1. *Another equality rule: isosceles triangles*. Another uses advanced fact 2.6.4 *Inscribed angles* and subtraction.

2.5.3 Draw a diagram

Some problems that sound very hard when they are asked in words are much easier if you draw a diagram. Here's one example:

> *Question: Regular pentagon AGHIF sits inside regular hexagon ABCDEF. The two shapes share base AF. Find the measure of angle BAG if it is less than 45°.*

This sounds very, very hard, but if you start by drawing the figure I drew above it doesn't seem quite as hard.

An IMLEM problem where this also helped me is

> *Question (IMLEM Meet #1, Oct. 2004): If the supplement of angle x is five times the complement of angle x how many degrees are in the measure of angle x?*

One way to solve this question is to write down and then solve the equation $180 - x = 5\,(90 - x)$. This isn't hard if you're good at algebra, but if you haven't had much algebra it is. My eight-year-old daughter definitely wasn't in the good-at-algebra category when we discussed this problem last year, but she managed to get the right answer pretty quickly anyway after I drew the diagram below:

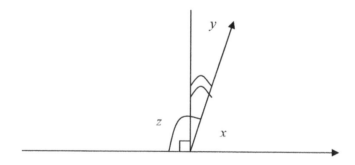

Looking at the figure makes you realize that if z is five times as big as y, then it must be that the right angle is four times as big as y. Hence $y = 90/4 = 22.5°$. You can then use the fact that x and y are complements to get $x = 90 - 22.5 = 67.5°$.

2.5.4 Draw in missing lines

Another trick that they use to make problems hard is to leave out lines that should be there and would make solving the problem much easier. The problem below is one example:

Question (IMLEM Meet #1, Oct. 2000): In the figure below rays AB and DE are parallel. Angle C is a right angle and angle B measures 52 degrees. Find the measure of angle D if it is less than 180 degrees.

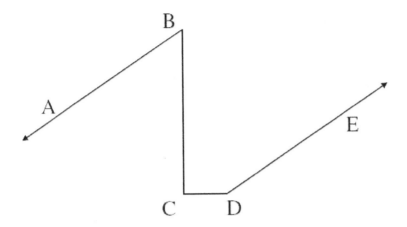

The reason why you can't think of anything to do when you first look at this problem is that they erased all the lines that would form the angles that need to be filled in to get to the answer. The first step in solving the problem is to extend segment BC further down and make ray DE into a line that goes further to the left until the two of them intersect. It's also useful to extend BC to go further up and to extend AB further to the right so you see all the angles they make where they intersect. Once you do this, there are lots and lots of things you can start filling in: you can use the supplementary-angles rule, you can use the adding-up-in-a-triangle rule, you can use corresponding-angles-on-parallel-lines-are-equal-rule, the opposite-angles-are-equal rule, etc. It's still not an easy problem because it takes several steps to get to the answer, but it's doable.

2.5.5 Save time by grouping angles

An aspect of the IMLEM math meets that can be very frustrating is that sometimes you'll feel that you could have much better, but for the fact they only give you 200 seconds to do each one. The next two subsections discuss tricks that sometimes let you get to the answer more quickly.

One trick that you can sometimes do is to save time by grouping angles together instead of trying to find the measure of every single angle. The problem below is one example.

Question (IMLEM Meet #1, Oct. 2001): In the figure below angles GRM and MRT are complementary. The measure of angle TRY is 127 degrees. Find the measure of angle GRY.

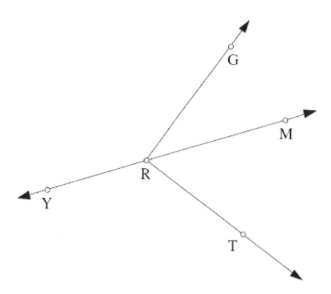

The brute force solution to this problem would be to first use the supplement rule to figure out MRY, and then use the complement rule to figure out MRG, and then add up several numbers using the adding-up-to-360 rule to find GRY.

The trick that would let you do this problem much more quickly is to notice that you don't need to know MRY and MRG to get the answer. An alternate way to do the adding-up-to-360 rule is to focus on **TRG** and note that TRY, TRG and GRY add up to 360. This gives 127 + 90 + GRY = 360.

My guess is that the reason they made angle TRY 127 degrees instead of 120 degrees in this problem is that they wanted all of the subtractions involved to be slow and frustrating. They probably wanted there to be a bigger advantage if you found the trick.

2.5.6 Save time by not doing all the subtractions

Another trick that comes in handy in some problems is to put off doing some of the subtractions until you're sure its necessary. Here's an example of where this technique saves a lot of frustrating work.

Question: In the figure below ABC and ADC are right angles. The measure of angle EAD is 127.72647 degrees. Find the measure of angle BCD.

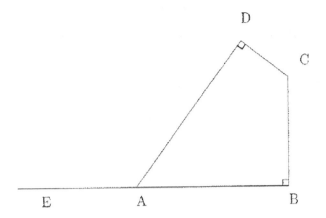

The first step in solving this problem is to use the supplementary-angle rule to find DAB. You could do the subtraction and find DAB = 52.27353 degrees, but you could also just write down DAB = 180 – 127.72647 and leave it at that. Why does this turn out to be a good idea?

Well, the next step in solving the problem is to use the adding-up-in-a-quadrilateral rule to find BCD.

If you spent all the time necessary to do the first subtraction, then this problem would look like 52.27353 + 90 + 90 + BCD = 360, and you'd have to do another annoying subtraction problem to get the answer.

If, on the other hand, you were lazy and stopped at DAB = 180 – 127.72647 the problem would look like (180 – 127.72647) + 90 + 90 + DAB = 360. The answer to this problem is obvious: DAB = 127.72647. The moral of this story: sometimes the lazy bird gets the worm.

2.6 Advanced Topics

I think that the facts I mentioned in sections 2.2-2.4 are all that one would have needed to know to solve all of the IMLEM problems from the past five years. There are, however, lots of other facts about angles that I would have assumed could come up too. I don't know why they haven't. It could be that there's a more detailed description of the contest topics somewhere that says they aren't covered. It could be just a coincidence that they haven't yet come up. In case it's the latter (and because it's always good to learn more) I decided to quickly cover a few more things.

2.6.1 Another equality rule: isosceles triangles

- Angles opposite equal sides in an isosceles triangle are equal.

One example of a problem where this comes in handy is:

Question: Line segment AB is perpendicular to line segment BC. D is a point on AB that is equidistant from A and C. If the measure of angle BCD is 40°, what is the measure of angle DCA.

A good way to start on this (and many other problems) is to draw a diagram. You'll note that DBC is a right triangle. Using the adding up rule for triangle DBC, the measure of angle CDB is 50°. ADC and CDB are complements, so the measure of ADC is 130°. Now, way that point D was defined in the question implies that triangle ADC is an isosceles triangle. Hence, angles DAC and DCA are equal. To get the angles to add up, each must be 25°.

You should always think of using the isosceles triangle rule when you're told that two line segments are the same length. One way in which people sometimes make this less obvious is to make up problems in which on vertex of a triangle is the center of a circle and the other two are points on the circle. All points on a circle are equidistant from the center, so this makes the triangle isosceles. You can also learn that a triangle is isosceles after you find that two angles are equal. It's good to make a note of this when it happens. Sometimes, the fact that two segments are equal can come in handy later on, e.g. if a third segment also ends up being equal in length to one of the first two.

2.6.2 Angle bisectors

An *angle bisector* is the set of all points equidistant from the two sides of an angle. The angle bisector divides an angle into two smaller angles. The measure of each is half of the measure of the full angle

2.6.3 Special triangles

An *equilateral* triangle is a triangle in which all sides and angles are equal. (You could call it a regular 3-gon if you wanted, but no one does.)

- Each angle in an equilateral triangle is 60°.

It's also good to know about two right triangles:

- In an isosceles right triangle the angles are 45°, 45°, and 90°. One way to recognize this triangle is that the length of the hypotenuse is $\sqrt{2}$ times the length of each leg. For example, if the legs of the triangles are 3 units then the hypotenuse is $3\sqrt{2}$ units. If the length of a leg is $\sqrt{2}$ units, then the length of the hypotenuse is 2 units.

- A second right triangle that comes up often is the 30-60-90 triangle. Again, you can recognize this triangle by the length of the sides. If the shortest side is x units, then the other leg will be $x\sqrt{3}$ units and the hypotenuse will be $2x$ units. I can imagine that they might ask a question where they give you the lengths of the sides and you have to recognize that it's a 30-60-90 triangle to find an angle you need to solve the problem.

2.6.4 Inscribed angles

Those of you who participated in IMLEM meet #4 last year may remember that one of the topics it covered were angles inscribed in circles. The definitions were:

- A *central angle* is an angle with its vertex at the center of the circle and its endpoints on the circle. In the figure on the left below below AOC is a central angle.

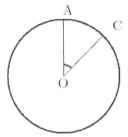 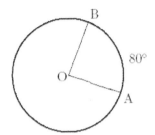

- An *arc* is the part of the circle between two points. In the figure on the right above, arc AB is the part of the circle between A and B.

- The *arc angle* is a measure of the relative size of an arc. Arc angle AB is defined to be the measure of central angle AOB. In the figure on the right above angle AOB and arc AB are both 80 degrees.

- An *inscribed angle* is an angle that has its vertex on the circle and both of its endpoints on the circle. In the figure on the left below ABC is an inscribed angle. In the figure on the right, ADB is an inscribed angle.

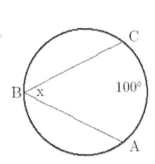 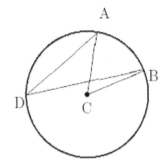

The main fact about inscribed angles you often need to solve problems is:

- The measure of an inscribed angle is one-half of the measure of the arc it intersects. In the figure on the left above, the measure of ABC is 50°. In the figure on the right above the measure of ADB is one half of the measure of ACB.

By now you may be wondering: "Why is he telling us about inscribed angles? There are no circles in meet #1!"

While this is true, a trick that you can use to figure out some problems is to draw in a circle. In particular, a fact you might want to keep in mind is that any regular *n*-gon can be inscribed in a circle. An example of a problem where this might be useful is the one below:

Question: In the figure below find the measure of angle GAI.

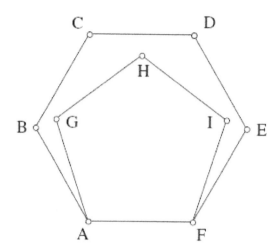

You could do this problem without knowing about inscribed angles by drawing in the line segment AI, figuring out the angle IAF by using the adding up rule in triangle IAF, and then subtracting GAI = GAF – IAF.

Thinking about inscribed angles gives you another way that would probably be quicker. Draw a the circumscribed circle around the pentagon and put a point O at the center. You may remember from before that each central angle connecting to a single side of the pentagon has a measure of 72° (one fifth of 360°). The central angle GOI is two of these angles together, so its measure is 144°. GAI is an inscribed angle in this circle, so its measure is one half of the measure of the corresponding central angle GOI. Hence, the measure of GAI is 72°.

2.6.5 Other units for measuring angles

In IMLEM angles always seem to have been measured in degrees. But just like lengths can be measured in inches or centimeters, and temperature can be measured in Fahrenheit or Celsius, there are also different scales that are sometimes used for measuring angles.

The most popular in advanced math is to measure angles in *radians*. 360° = 2π radians. A right angle is $\pi/2$ radians. A 30° angle is $\pi/6$ radians. For the kinds of questions they typically ask in IMLEM meet 1 this sounds like a crazy system to use. But there are other types of problems where it is more convenient. I won't tell you about any of them. I just figured I'd mention radians in case they happened to be included on some meet.

If you have a good scientific calculator you may have noticed that it has a mysterious DRG button. This button lets you shift between three methods for measuring angles. D stands for degrees. R stands for radians. G stands for *grads*. In the grad system a right angle measures 100 grads. You might think that this is a good idea for the same reason that the metric system is a good idea. It turns out, however, that using grads as the unit of measurement hardly ever makes problems easier to work with and mathematicians never do it.

Category 3 – Number Theory

The number theory category in meet #1 focuses on prime numbers. It includes prime factorization, divisibility rules, and some counting techniques

3.1 Prime Numbers

The first thing you need to know for this category is what the word *factor* means.

- We say that the number n is a *factor* of the number m if there exists a third number k such that $n \times k = m$.

Example: The factors of 44 are 1, 2, 4, 11, 22, and 44. You can check that each of these are factors by thinking of how to get 44 by multiplying the number by another whole number. For example, the reason that 2 is a factor is that $2 \times 22 = 44$. Other numbers like 5 are not factors because there's no way to multiply 5 by an integer and get 44.

The *natural numbers* are the standard numbers you learned in kindergarten (or whenever it was that they first taught you numbers but didn't want to confuse you by mentioning zero): 1, 2, 3, 4, 5 … The most important thing to know for category 3 of meet 1 is the definition of a prime number:

- A *prime number* is a natural number that has exactly two factors: one and itself.

For example, two is prime because it has two factors: 1 and 2. Seventeen is prime because the only way to get 17 as the answer to a multiplication problem is $1 \times 17 = 17$. Twenty one is not prime because are different ways to multiply two numbers together and get 21. You can do $1 \times 21 = 21$ or $3 \times 7 = 21$.

A tricky part of this definition is that by saying "exactly two" I've ruled out the number one. It does not meet the definition of a prime number because it has only one factor. This is good because the number one is not prime.

You can do problems in this category much more quickly if you can look at a number and immediately know whether it is prime, rather than having to try to figure it out. Here's list of the first fifty prime numbers. There's not really any point in memorizing the list in the sense of being able to repeat back the prime numbers in order. But it is useful to look at it enough so that you mostly know which numbers are prime, especially the ones that are less than 50.

- 2, 3, 5, 7, 11, 13, 17, 19, 23, 29, 31, 37, 41, 43, 47, 53, 59, 61, 67, 71, 73, 79, 83, 89, 97, 101, 103, 107, 109, 113, 127, 131, 137, 139, 149, 151, 157, 163, 167, 173, 179, 181, 191, 193, 197, 199, 211, 223, 227, 229.

Most meets have a question that you can answer if you just know what a prime number is and have a pretty good idea of what numbers are on this list. For example,

Question: Numbers a and b are said to be a pair of twin primes if they are both prime and differ by two. Find the smallest pair of twin primes greater than 20.

Question: Jim-Bob added two different primes together and got 38. What were they?

Question: Two primes both less than fifty have a difference of six. What is the largest possible value that the sum of these primes could have?

The answers to these questions are 29 and 31, 7 and 31, and **88**.

One additional term you will see on the meets is

- A whole number is *composite* if it is not prime (and greater than one).

Twenty-one is composite because $3 \times 7 = 21$. Fifty four is composite because is can be factored in lots of ways, e.g. $6 \times 9 = 54$ and $2 \times 27 = 54$.

3.2 Prime Factorization

3.2.1 The Fundamental Theorem

The next most important thing to after the definition of a prime number is

- *Fundamental Theorem of Arithmetic*: Every whole number greater than one can be uniquely written as a product of primes.

For example we have,

$3 = 3$
$4 = 2 \times 2$
$6 = 2 \times 3$
$33 = 3 \times 11$
$91 = 7 \times 13$
$30 = 2 \times 3 \times 5$
$44 = 2 \times 2 \times 11$
$54 = 2 \times 3 \times 3 \times 3$

$660 = 2 \times 2 \times 3 \times 5 \times 11$

Note that we sometimes need to use the same prime number several times in a prime factorization. For example, we did this in $4 = 2 \times 2$.

3.2.2 Exponential notation

Something you probably learned in elementary school is that another way to write $4 = 2 \times 2$ is with exponents: $4 = 2^2$. Some prime factorizations written with exponents are:

$44 = 2^2 \times 11$
$54 = 2 \times 3^3$
$540 = 2^2 \times 3^3 \times 5$

People often replace the $\times$ signs in these expressions by a small dot or leave it out entirely. For example, you may see $44 = 2^2 \times 11$ written as $44 = 2^2 \cdot 11$ or $44 = 2^2 \, 11$. IMLEM sometimes asks you about prime factorizations by asking you about what it looks like when you factor it at put it in this form.

> *Question: Suppose that $540 = 2^a \cdot 3^b \cdot 5^c \cdot 7^d \cdot 11^e$ (and that a, b, c, d, and e are integers). Find $a + b + c + d + e$.*

I noted above that the factorization of 540 is $540 = 2^2 \cdot 3^3 \cdot 5$. If you remembered this (or factored the number correctly yourself) you probably realized that a=2 and b=3, but may not have figured out what c, d, and e were. There are a couple tricks here: One is that 5 can also be written as 5^1. Another is n^0 is always equal to 1. Hence, another way to write the prime factorization is $540 = 2^2 \cdot 3^3 \cdot 5^1 \cdot 7^0 \cdot 11^0$. The answer is $2 + 3 + 1 + 0 + 0 = 6$.

Here's another example (in the past they've left off the part of the question saying that a, b, c, d, and e are integers even they should have said it).

> *Question: Suppose that $2000 = 2^a \cdot 3^b \cdot 5^c \cdot 7^d \cdot 11^e$. Find the maximum of a, b, c, d and e.*

3.2.3 Finding prime factorizations

The easiest way to find prime factorizations is usually just to keep looking for one factor at a time and adding them to what you've found. For example,

$84 = 2 \times 42$
$\quad = 2 \times 2 \times 21$
$\quad = 2 \times 2 \times 3 \times 7$

When you're doing this it helps to use rules you know for divisibility. For example, in the problem above as long as the number was even I knew I could keep dividing by two. I'll discuss many more divisibility rules in section 3.4.

In some cases, you can do problems more quickly by taking advantage of things you know. For example, if I was asked to factor 3500 the way I'd do it is

$3500 = 35 \times 100$
$\quad = (5 \times 7) \times (10 \times 10)$
$\quad = (5 \times 7) \times (2 \times 5 \times 2 \times 5)$
$\quad = 2^2 \times 5^3 \times 7$

3.3 Counting and Summing Factors

Several IMLEM questions have asked how many factors a number has or what the sum of all of the factors of a number is. For example,

Question: How many factors does 200 have?

Question: List all of the factors of 200.

Question: What is the sum of all of the factors of 200.

The first way that you might think of to solve problems like this is to go through all of the numbers between 1 and 200 and check whether they're factors. One, $1 \times 200 = 200$, so yes. Two, $2 \times 100 = 200$, so yes. Three, no. Four, yes. Five, yes. And so on. **Don't do it!** IMLEM only gives you two hundred seconds to do each problem. Checking 200 numbers in 200 seconds is not a viable plan.

Instead, the way to do problems like this is to think about prime factorizations. Here's how.

3.3.1 Counting and listing factors: factor tables

How could you know right away that 17 is not a factor of 200? You could multiply things out and find $11 \times 17 = 187$ and $12 \times 17 = 204$, but there's a better way:

The prime factorization of two hundred is $200 = 2^3 \cdot 5^2$. If 17 were a factor of 200, then there would have to be some expression like $200 = 17 \times n$. If this were true, however, then you could find the prime factorization of 200 by starting with this expression, then factoring n into its factors, and so on.

$200 = 17 \times n$
$\quad = 17 \times (m \times r)$
$\quad = 17 \times (m \times s \times t)$
$\quad = \ldots$

Note though that regardless of what happens when you factor n, you'd always end up with a 17 in the prime factorization. We know that the prime factorization of 200 has nothing other than 2's and 5's in it, so this couldn't be right.

What can the factors of 200 be? We know 17 isn't a factor, nor is many multiple of 17. Nor is 3, nor any multiple of 3. Nor 7, nor any multiple of 7, etc. What's left? What's left is numbers that have only 2's and 5's in their prime factorization. More specifically, **all factors of 200 are of the form $2^a \cdot 5^b$ with $a \leq 3$ and $b \leq 2$.**

We can easily list all of the factors by making a factor table. It looks like this.

	1	2	$2^2=4$	$2^3=8$
1 |	1	2	4	8
5 |	5	10	20	40
$5^2=25$ |	25	50	100	200

Making a table like this lets us answer the two questions I started this section with.

Answer: 200 has 12 factors

Answer: The factors of 200 are 1, 2, 4, 5, 8, 10, 20, 25, 40, 50, 100, and 200.

What do you do when a number has more than two distinct prime factors? For example, what are the factors of 600? The prime factorization of 600 is $600 = 2^3 \cdot 3 \cdot 5^2$. What you do in a case like this is just to make multiple tables. The table on the left is all factors that aren't a multiple of 3. The table on the right is all factors that are. There are 24 factors in all.

1 |	1	2	4	8
1 |	1	2	4	8
5 |	5	10	20	40
25 |	25	50	100	200

3 |	1	2	4	8
1 |	3	6	12	24
5 |	15	30	60	120
25 |	75	150	300	600

What if a number has four factors? An example is $4200 = 2^3 \cdot 3 \cdot 5^2 \cdot 7$. There are 48 factors:

1 |	1	2	4	8
1 |	1	2	4	8
5 |	5	10	20	40
25 |	25	50	100	200

3 |	1	2	4	8
1 |	3	6	12	24
5 |	15	30	60	120
25 |	75	150	300	600

7 |	1	2	4	8
1 |	7	14	28	56
5 |	35	70	140	280
25 |	175	350	700	1400

3×7 |	1	2	4	8
1 |	21	42	84	168
5 |	105	210	420	840
25 |	525	1050	2100	4200

Of course, they'd never ask you to list all these factors during a meet. It'd be hard to write all of them down in the time they give you. More importantly, there's no way that the teachers would want to correct such a long list! They certainly have, however, asked questions like

Question: How many factors does 4200 have?

Question: For how many positive integers n is 4200/n an integer.

The two questions are two ways to ask the same question. For this reason, they have the same answer: 48. The reason why this question is reasonable to ask is that you don't need to fill in the factor table to answer the question. You just have to think about what the factor table(s) would look like and multiply (number of rows) × (number of columns) × (number of tables) to figure out how many numbers would be in them. We can write this as a formula:

- If the prime factorization of n is $n = 2^a \cdot 3^b \cdot 5^c \cdot 7^d \cdot 11^e \cdot 13^f \cdot \ldots$, then the number of factors of n is $(a+1)(b+1)(c+1)(d+1)(e+1)(f+1)\ldots$

This is one of the most important formulas to learn for the meet. One example of an IMLEM problem where knowing this makes the question pretty easy is

Question (IMLEM Meet #1, October 200): For how many positive integral values of n will 168/n be a whole number?

The prime factorization of 168 is $168 = 2^3 \cdot 3 \cdot 7$. The answer is $(3+1)(1+1)(1+1) = 16$.

A question that would have seemed much harder than this one if you didn't know the formula, but is not much harder if you know the formula is

Question: How many factors does 1,000,000 have?

The person who makes up the IMLEM tests seems to try to make the third question on each round and the questions in the team round harder than the rest of the questions. When he's making up one of these extra-hard questions he might also ask something that's similar to the questions above, but takes some thinking. For example,

Harder Question: How many of the factors of 168 are even numbers?

The answer is 12. One way to think about is to think that the even factors will be all factors of the form $2^a \cdot 3^b \cdot 7^d$ with a≤3, b≤1, c≤1 **and** a≥1. This gives 3 choices for a, 2 choices for b, and 2 choices for d, so the answer is $3 \times 2 \times 2 = 12$. Another way to get the answer is to think about the factor tables. If you were put 1, 2, 4, and 8 along the top of the tables, then the only odd factors in each table are the numbers in the column below the number 1. There are two rows in each factor table, so each has just two odd numbers in it. There are two tables, so there are $2 \times 2 = 4$ odd factors. There are 16 factors in total, so 12 must be even.

3.3.2 Summing factors

Another question they like to ask sometimes is for the sum of all the factors of a number. For example,

Question: Find the sum of all of the factors of 200.

You could answer this by making a factor table, finding that the factors are 1, 2, 4, 5, 8, 10, 20, 25, 40, 50, 100, and 200, and then adding all these together. This is doable, but slow. It turns out that there's a formula you can use to get the answer much more quickly.

Here's the formula written mathematically:

- If the prime factorization of n is $n = 2^a \cdot 3^b \cdot 5^c \cdot 7^d \cdot 11^e \cdot 13^f \cdots$, then the sum of the factors of n is $(1 + 2 + 2^2 + \ldots + 2^a)(1 + 3 + 3^2 + \ldots + 3^b)(1 + 5 + \ldots) \ldots$

For example, the prime factorization of 200 is $n = 2^3 \cdot 5^2$, so the sum of its factors is $(1 + 2 + 4 + 8)(1 + 5 + 25) = 15 \times 31 = 465$.

Another way to think about the formula that may make it easier to remember is to think about it in terms of factor tables. Recall that the factor table for 200 is

```
                1+2+4+8=15
            |   1    2    4    8
          --|------------------------
          1 |   1    2    4    8
1+5+25=31 5 |   5   10   20   40
         25 |  25   50  100  200
```

What the formula says you should do in this case is to add the numbers written along the top, 1+2+4+8=15, add the numbers written along the side, 1+5+25=31, and then multiply these two numbers together, $15 \times 31 = 465$.

When you need more than one factor table to list all the factors, the formula says you should also add up the numbers you wrote in the upper left boxes and multiply these. For example, when I listed the factors of 600 I wrote 1 in the upper left part of the first table and 3 in the upper left part of the second factor table. The sum of all the factors of 600 is $(1 + 2 + 4 + 8)(1 + 5 + 25)(1 + 3) = 15 \times 31 \times 4 = 1860$.

Why does this formula work?

Think about what happens when you multiply $(a + b + c)(d + e)$. The standard rules for doing this tell you $(a + b + c)(d + e) = (ad + ae) + (bd + be) + (cd + ce)$. If we expand the product $(1 + 5 + 5^2)(1 + 2 + 2^2 + 2^3)$ in this way we find

$$(1+5+5^2)(1+2+2^2+2^3) = (1 + 2 + 2^2 + 2^3) + (1 \cdot 5 + 2 \cdot 5 + 2^2 \cdot 5 + 2^3 \cdot 5) + (1 \cdot 5^2 + 2 \cdot 5^2 + 2^2 \cdot 5^2 + 2^3 \cdot 5^2)$$

The numbers on the right side of this expression are all of the numbers we'd put in the factor table.

Another equivalent version of the formula above is

- If the prime factorization of n is $n = 2^a \cdot 3^b \cdot 5^c \cdot 7^d \cdot 11^e \cdot 13^f \cdot \ldots$, then the sum of the factors of n is

$$\frac{(2^{a+1} - 1)(3^{b+1} - 1)(5^{c+1} - 1) \ldots}{(2-1)(3-1)(5-1) \ldots}$$

You could try to learn this one instead if you think its easier. If you know algebra it's an interesting problem to figure out why the two formulas give the same answer.

Here's a couple of extra questions you could do for practice if you wanted:

Question: What is the sum of all of the factors of 66?

Question: What is the sum of all of the factors of 64?

Question: What is the sum of all of the odd factors of 66?

3.4 Divisibility Rules

When you're trying to find the prime factorization of a number it helps to be able to figure out things like whether the number is a multiple of 3 very quickly. Several IMLEM questions also ask you directly whether a number is a multiple of another. For example,

Question: For what values of A is the five digit number 3124A a multiple of 6.

In both situations it helps to know divisibility rules.

3.4.1 Famous divisibility rules

The three best known divisibility rules are:

- A number is a multiple of 10 if it ends in a zero.

- A number is a multiple of 2 if the last digit is 0, 2, 4, 6, or 8.

- A number is a multiple of 5 if the last digit is 0 or 5.

Another one that some of you may have seen is

39

- A number is a multiple of 3 if the sum of the digits is a multiple of 3.

For example, if you wanted to figure out whether 31,245 is a multiple of 3 you would add the digits, $3+1+2+4+5 =15$. This is a multiple of 3, so you know that 31,245 is a multiple of 3. (If you don't believe me just multiply out $3 \times 10,415$.)

A nice little trick to keep in mind in some problems is that you can use this formula repeatedly. For example, is 9988877665 a multiple of 3? Well, $9+9+8+8+8+7+7+6+6+5=73$. If you don't remember whether 73 is a multiple of 3, just apply the rule again. $7+3=10$. Most of you know that this is not a multiple of 3. If you didn't though, you could use the rule again: $1+0=1$. Hopefully, at this point you'd recognize that 1 is not a multiple of 3.

3.4.2 Less famous divisibility rules

The rule for recognizing an even number is related to rules for other powers of two:

- A number is a multiple of 2 if the last digit is an even number.

- A number is a multiple of 4 if the number formed by the last two digits is a multiple of 4. For example, 31,220 is a multiple of 4 because 20 is a multiple of 4 and 310 is not a multiple of 4 because 10 is not a multiple of 4.

- A number is a multiple of 8 if the number formed by the last three digits is a multiple of 8.

- A number is a multiple of 16 if the number formed by the last four digits is a multiple of 16.

These rules get hard to apply for big powers of 2, so people have also invented other rules that can be easier to apply. I'll state them as they apply to five digit numbers $abcde$, but they apply to all numbers in the obvious way.

- The number $abcde$ is a multiple of 2 if e is multiple of 2.

- The number $abcde$ is a multiple of 4 if $2d + e$ is multiple of 4.

- The number $abcde$ is a multiple of 8 if $4c + 2d + e$ is multiple of 8.

- The number $abcde$ is a multiple of 16 if $8b + 4c + 10d + e$ is multiple of 8. (Sorry, I know that this ruins the pattern, but it's not my fault. That's just the way it is.)

The rule for divisibility by 9 is similar to the rule for divisibility for 3.

- A number is a multiple of 3 if the sum of the digits is a multiple of 3.

- A number is a multiple of 9 if the sum of the digits is a multiple of 9.

The rule for divisibility for 11 looks a little like this and is kind of neat.

- The number *abcde* is a multiple of 11 if the alternating sum/difference of the digits is a multiple of 11, i.e. if $e - d + c - b + a$ is a multiple of 11. For example, the number 14641 is a multiple of 11: $1 + 4 - 6 + 4 - 1 = 0$

The powers of 5 have pretty easy rules to remember:

- A number is a multiple of 5 if the last digit is a multiple of 5.

- A number is a multiple of 25 if the number formed by the last two digits is a multiple of 25.

Why all these rules work will be much easier to understand after you've read the meet #4 arithmetic section on modular arithmetic. After reading that section you'll be able to make up and explain all kinds of rules including

- The number *abcde* is a multiple of 17 if and only if $e + 10d - 2c - 20b + 4a$ is a multiple of 17.

For now you'll just have to trust me (or jump ahead to the modular arithmetic section).

3.4.3 Divisibility by composite numbers

If you look back at all of the divisibility rules I mentioned you'll note that they are all about divisibility by a prime number or a power of a prime number. Most IMLEM questions on divisibility seem to ask about divisibility by a composite number or about divisibility by multiple primes. For example, they might ask

Question: For what values of A is the five digit number 3124A a multiple of 6.

Or

Question: For what values of A is the five digit number 967A0 divisible by 2, 3, 4, and 6?

The key to answering all questions like this is:

- A number n is divisible by $2^a \cdot 3^b \cdot 5^c \cdot 7^d \cdot 11^e \cdot 13^f \cdot \ldots$ if and only if it is divisible by 2^a **and** divisible by 3^b **and** divisible by 5^c **and** ...

For example, the number 3124A is divisible by 6 if it is a multiple of 2 and a multiple of 3. It is a multiple of 2 if A is 0, 2, 4, 6, or 8. It is a multiple of 3 if 3+1+2+4+A is a multiple of 3. This is true if A is 2, 5, or 8. Hence, the answer to the first question above is that 3124A is a multiple of 6 for A=2 and for A=8.

41

To check whether 967A0 is divisible by 2, 3, 4, and 6, we just need to check whether it is a multiple of 2^2=4 and 3. It is a multiple of 4 if $2 \cdot A + 0$ is a multiple of 4. This is true whenever A is an even number. 967A0 is a multiple of 3 if 9+6+7+A+0=22+A is a multiple of 3. This happens for A=2, 5, and 8. Hence, 967A0 is a multiple of 3 and 4 if A=2 or A=8.

3.5 Problem Solving Tips

A final type of problem they seem to like to ask occasionally is questions where they give multiple clues about a number or set of numbers and ask you to find the number(s). For example,

> *Question (IMLEM Meet #1, October 2002): Find the value of n if*
> *n is a natural number,*
> *n is less than 100,*
> *n is the product of two primes,*
> *the sum of the digits of n is 10, and*
> *the positive difference between the two prime factors of n is a multiple of 13.*

> *Question (IMLEM Meet #1, October 2000): Find the sum of all positive integers less than 1000 that are both perfect squares and perfect cubes.*

Here are a couple tips for solving such problems.

3.5.1 Focus first on whatever lets you cut down to a small set

In the question with multiple clues, a clue that immediately cuts you down to a small number of possibilities is the fourth clue: the sum of the digits of *n* is 10. This lets you know that the number is 19, 28, 37, 46, 55, 64, 73, 82, and 91. Once you've gotten to this point, you can just check each of these numbers to which satisfies the other properties: (1) a product of two primes; and (2) the positive difference between the two prime factors is a multiple of 13.

19, 37, and 73 are out because they're prime. 28, 64 are not the product of two primes. This leaves 46 (= 2 ×23), 55 (= 5 × 11), 82 (=2 × 41), and 91=(7 ×13). The two prime factors differ by a multiple of 13 for *n* = 82.

3.5.2 Focus on possible factors rather than on possible numbers

The problems in this section are mostly about prime factorization. That means that most of them can be solved more easily by thinking about prime factorization. Consider, for example, the second problem I mentioned above:

> *Question (IMLEM Meet #1, October 2000): Find the sum of all positive integers less than 1000 that are both perfect squares and perfect cubes.*

You could try to solve this by going through all numbers less than 1000 and checking whether they work, but that would take a very long time. Instead, think about what it means for a number to be both a perfect square and a perfect cube.

The number $n = 2^a \cdot 3^b \cdot 5^c \cdot 7^d \cdot 11^e \cdot 13^f \cdot \ldots$ is a perfect square if a, b, c, d, e, f, … are all even numbers. It is a perfect cube if a, b, c, d, e, f, … are all multiple of 3. These will both be true if and only if all of these exponents are a multiple of 6, i.e. if n is a perfect 6th power. The 6th powers that are less than 1000 are $1^6 = 1$, $2^6 = 64$, and $3^6 = 729$. Hence the answer is $1 + 64 + 729 = 794$.

Category 4 – Arithmetic

The arithmetic category in meet #1 covers order of operations and statistics. In some ways it's one of the easiest topics they have in IMLEM: you only need to know a few definitions; the answers are never complicated to work out; and knowing a few tricks makes some of the problems pretty easy. The one thing that can make the category difficult is that you often need to do a lot of multiplying, dividing, adding, subtracting, and rounding off in a limited amount of time. This makes it a good category to choose for people who are good at doing these things in their head.

4.1 Order of Operations

Order of operations questions just ask you to do a sequence of standard arithmetic operations. The one thing that's tricky about them is that they leave out the parenthesis that would remind you in which order to do things.

> *Question: Find the value of the expression: $100 - 3(42 - 5^2 \times 3 + 4^3 \div 8)$.*

The rules for evaluating expressions like this are:

- Evaluate expressions in parentheses as soon as you can

  e.g. $2 (3 + 5) = 16$ and $2^{(5-3)} = 2^2 = 4$

- Do exponents (including square roots) first, then multiplications and divisions, then additions and subtractions.

  e.g. $44 - 5 \times 2^3 = 44 - 5 \times 8 = 44 - 40 = 4$.

- Within a category (e.g. when doing multiple additions/subtractions) do things from left to right.

  e.g. $6 - 3 - 2 = 3 - 2 = 1$ and $12 \div 6 \times 2 = 2 \times 2 = 4$.

In some schools they use the mnemonic PE^{MA}_{DS} to help remember this: Parentheses, Exponents, Multiplication & Division, Addition & Subtraction. Applying these rules you should find that the answer to the question I started with is 175.

There really isn't much more to know when doing problems in this category. In a few cases problems could me done more quickly using one of the two tricks below:

- Look for terms you can group together.

For example, if asked to evaluate $333 \times (163 - 88) - 45 \times 4 - 329 \times (163 - 88)$, you should recognize that this is equal to $(333 - 329) \times (163 - 88) - 45 \times 4$. This saves you the effort of multiplying 333×75, and also makes you realize that you only need to do the subtraction once.

- Remember that $(a^2 - b^2) = (a - b)(a + b)$.

    For example, it's easiest to compute $(27^2 - 23^2)$ via $(27^2 - 23^2) = (27 - 23)(27 + 23) = 200$.

One other thing you probably know without knowing it is that the way we write exponents and fractions can imply that you should treat them as if there were parentheses there even though we don't write the parentheses. For example, it's understood that 2^{5-3} means $2^{(5-3)}$ and that $\frac{11-4}{7}$ means $(11 - 4)/7$.

4.2 Statistics

The statistics questions usually give you a set of numbers and ask you to compute a few things about them.

The sets are presented in a few different ways. Sometimes they give you a sorted list of numbers, e.g. 90, 90, 90, 90, 90, 95, 95, 95, 100. Sometimes they give you an unsorted list, e.g. 90, 95, 95, 90, 90, 90, 100, 95, 90. Sometimes they'll describe the list, e.g. 5 students scored 90, 3 students scored 95, and 1 student scored 100. Sometimes they'll describe it by giving a histogram:

```
            X
            X
            X           X
            X           X
            X           X           X
    _____
     85     90          95          100
           Students' Scores on Test
```

You just need to remember that all of these are equivalent ways to describe a set.

They seem to mostly ask you to compute four things from data:

4.2.1 Range

The *range* of a set of data is the difference between the largest number of a set and the smallest. For example, the range of the above dataset is $100 - 90 = 10$.

4.2.2 Mode

45

The *mode* is the number that occurs the largest number of times. For example, the mode of the above dataset is 90. The mode of 25, 26, 26, 27, 27, 27, 30, 30 is 27. This is only well-defined if there is a single number that occurs more times than any other number. They shouldn't ask you problems in which this isn't true.

4.2.3 Median

The median of a set is a number such that half of the elements are larger (or the same) and half are smaller (or the same). For example, the median of 25, 26, 30, 40 and 50 is 30.

When the set has an even number of numbers (and the two in the middle are different) IMLEM defines the median to be the average of the two middle scores. For example, the median of 5, 10, 15, and 25 is 12.5.

When numbers are repeated in a dataset, be sure to count them multiple times. For example, the median of 20, 20, 20, 22, 24, 26, and 30 is 22. Another example is that the median of the dataset described at the start of this section is 90: the four largest elements are 95, 95, 95, and 100. The four smallest are 90, 90, 90, and 90. The one number left over in between these is 90. Another example is:

> *Question: On the last IMLEM math meet twenty two teams scored 0 on the team round, fourteen scored 6, ten scored 12, two scored 18, and one scored 24. What was the median score.*

By giving you a list of five different scores they are trying to get you to say 12. Don't. The correct answer is 6. There are 49 teams in all. The 25th lowest score is a 6. The 25th highest score is a 6. Another example would be

> *Question: In the first IMLEM math meet two Bigelow students scored zero, one scored 4, one got 6, one got 8, three got 10, one had 14, and one had 18. What was Bigelow's median score?*

The answer to this is 9. The two scores in the middle are the 8 and one of the 10's.

One way to find the median of a large set is to count how many elements there are and then count half that many elements down from the top. If the set has 27 elements then the median is the 14th largest (and 14th smallest). If the set has 40 elements than the median is the average of the 20th and 21st largest.

Another good way to do many problems is to not bother counting and just start crossing equal numbers off both ends of the list. For example:

Question: The histogram below shows students' scores on a math test. What was the median score?

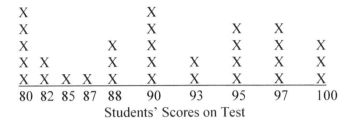

Students' Scores on Test

You could do this by counting up all the students, finding that there are 30, and then counting down from the top to find the 15th and 16th highest scores to take the average. For me though, I think it's quicker to use the crossing off method. To do this I first notice that there are seven students at 80 or 82 and seven at 97 or 100, so I can cross out all of these. The next highest scores are four 95's. I cross these out and at the same time cross out the four lowest remaining scores: 85, 87, and two of the 88's. I then cross out the two 93's and two lowest scores remaining: the last 88 and one of the 90's. I'm left with nothing but 90's, so the answer is 90.

4.2.4 Mean or Average

The most common questions on IMLEM meet #1 have been questions about means (also known as averages).

The *mean* of a set of numbers $\{x_1, x_2, \ldots, x_N\}$ is $(x_1 + x_2 + \ldots + x_N) / N$. For example, the mean of 10, 10, 20 and 60 is $(10 + 10 + 20 + 60)/4 = 25$.

Again, remember that you need to count each number as many times as it appears. For example, in the question below you need to keep this in mind both when computing the median and when computing the mean.

Question (IMLEM meet #1, Oct. 2003): The prize money at a raffle was given out as follows:
   20 people received $5
   10 people received $10
   5 people received $20
   2 people received $50
   1 person received $100
   1 person received $500
   1 person received $1000
How much greater was the mean of the prize money than the median?

The median is $(5 + 10)/2 = 7.5$. The average is $(20{\times}5 + 10{\times}10 + 5{\times}20 + 2{\times}50 + 1{\times}100 + 1{\times}500 + 1{\times}1000) / 40 = 2000/40 = 50$. The answer is $42.50. (Remember to put it in dollars and cents or they'll mark it wrong.)

Sometimes a question will just ask you to compute the mean of a set of numbers. More often though they ask you something a little more complicated. Here are some tricks to keep in mind that can make the problems a little easier:

- Think about what the *sum* of all the numbers must be.

  Question (IMLEM Meet #1, Oct. 2000) Sly knows that his quiz scores were 92, 85, 96, 82, 88, and 91 for the first quarter, but he can't remember what he earned on the test. His teacher told him that his combined average for the quizzes and the test is exactly 90 and that she counts the test as two quizzes. What score did Sly get on his test?

The answer is 93. If he had an average of 90 on eight scores (six quizzes plus a test that counts double) then he must have earned $90 \times 8 = 720$ points in total. On the six quizzes he earned $92 + 85 + 96 + 82 + 88 + 91 = 534$ points. Hence he must have earned $720 - 534 = 186$ points on the test. $186/2 = 93$.

- Use the average of averages rule. If the average of m numbers is a and the average of n numbers is b then the average of all $(m + n)$ numbers is $(ma + nb)/(m + n)$.

  Question (IMLEM Meet #1, Oct. 2000) Old McDonald has 7 cats and 4 dogs. The average weight of the cats is 12 pounds and the average weight of the dogs is 33 pounds. What is the average weight of Old McDonald's eleven animals? Round your answer to the nearest tenth of a pound.

The answer is 19.6. The answer comes directly from applying the above formula: $(7 \times 12 + 4 \times 33)/11 = (84 + 132)/11 = 216/11 = 20 - 4/11 \approx 19.6$.

- Use ahead/behind counting: if the average of the set $\{x_1, x_2, ..., x_N\}$ is a then it must be that $(x_1 - a) + (x_2 - a) + ... + (x_N - a) = 0$

This trick seems to help on almost every IMLEM problem on means. For example, the way I actually did the problem about Sly's tests and quizzes in my head when I first saw it was to say: $92 - 90 = 2$, so after one quiz he was 2 points ahead; $85 - 90 = -5$, so after the second quiz he lost 5 points and fell to 3 points behind; $96 - 90 = 6$, so after the third quiz he was 3 points ahead; $82 - 90 = -8$ left him 5 points behind; $88 - 90 = -2$ made him 7 points behind; and $91 - 90 = 1$ left him 6 points behind. He therefore must have caught up six points on the final test. It counted double so he could do this by scoring 93.

This method also lets you find the averages of sets of big numbers using a kind of guess-and-check strategy. For example,

Question: Find the average of 3630, 3669, 3698, 3703, and 3725.

You could do this by adding up the numbers and dividing by five, here's another way. Start off with a nice even number as a guess. In this problem I'd guess 3700. Then start counting up how far ahead or behind the guess you are: -70 + -31 + -2 + 3 + 25 = -75. This means that the true average must be 75/5 less than the number we guessed. The answer is 3685.

As a final example here's a problem similar to the one earlier where I used a method that's a hybrid of these two:

Question: (IMLEM meet #1, Oct. 2005) So far this quarter Gil has taken 6 quizzes and 2 tests and there is one more test to take. His scores on the quizzes are 85, 83, 98, 93, 86, and 87. His scores on the tests are 91 and 87. In computing the average Gil's teacher counts each test as two quizzes. He will round an average of 89.5 up to 90, which is an A-. What is the lowest score that Gil can get on the final test of the quarter and still have an average of 89.5?

The way I'd do this is to say that I know that he needs to end up at most 6 points behind a 90 average on the twelve grades (six quizzes plus three tests which each count double.) So far what he'd done relative to getting 90 on everything is -5 + -7 + 8 + 3 + -4 + -3 + (2 × 1) + (2 × -3). This leaves him at -12, or 12 points behind what he needs for a 90 average. To get to -6 he'll need to catch up 6 points on the final test, which he can do with a 93.

Category 5 – Algebra

The algebra category in meet #1 covers several related topics: simplifying expressions; evaluating expressions; solving equations in one unknown; identities; and made-up operations. It's not a hard category if you know algebra, although it's similar to category 4 in that they sometimes ask questions that require doing a lot of adding, subtracting, multiplying, and dividing per unit time.

Math team is a team sport. The algebra category is the place where this most often comes up – some teams don't have six people who are good at algebra. When this happens teams sometimes ask some of their younger players to sacrifice for the good of the team and do the algebra category even though they don't know algebra. Meet 1 can be an especially good time for this – it doesn't cover a lot of material, so a student who's good at calculations will have a good chance at getting a 2 or 4 if they study, even if they don't know algebra.

I don't have enough space in this packet to teach algebra to people who don't know it. I will try to review the types of problems that will be on the test. In some cases there are tricks that have appeared over and over that let you answer the problems more quickly. I'll try to point them out.

5.1 Simplifying expressions

A typical simplifying expressions problem is

Question: Simplify $6(2x + 1) + 5x - 3(2x - 7) + 14 - 5x$.

There isn't much to say about how to do this one. Basically, you just need to multiply things out and collect the terms together:

$$6(2x + 1) + 5x - 3(2x - 7) + 14 - 5x = 12x + 6 + 5x - 6x + 21 + 14 - 5x$$
$$= 6x + 41.$$

- One method that some people find useful to avoid making mistakes in problems like this is to arrange the terms in columns and then add down the columns. For example, you can write out the above expression as

$$
\begin{array}{rl}
 & 12x + 6 \\
+ & 5x \\
+ & -6x + 21 \\
+ & \underline{-5x + 14} \\
= & 6x + 41
\end{array}
$$

- A trick that comes up a lot is that you should always look for terms that can be grouped together to simplify the calculation. For example, suppose you are asked

Question: Simplify $237 (12x - 9) + 3x + 5 - 235 (12x - 9)$

If you try to do this by multiplying everything out it could take a very long time and you'd likely make an arithmetic mistake. The easier way to do this to recognize that you should group the first and last term together and simplify it that way:

$$237\,(12x - 9) + 3x + 5 - 235\,(12x - 9) \quad \begin{aligned} &= (237 - 235)\,(12x - 9) + 3x + 5 \\ &= 2\,(12x - 9) + 3x + 5 \\ &= 24x - 18 + 3x + 5 = 27x - 13. \end{aligned}$$

5.2 Evaluating expressions

Evaluating expressions problems just ask you to plug a value for a variable into an expression and evaluate it. For example, a really easy question would be:

Question: Evaluate $5x + y$ for $x=3$ and $y=2$.

The answer to this is $5 \times 3 + 2 = 17$. Real IMLEM problems require a few more calculations than this and every single one can be done more simply with the same trick:

- If plugging in will be a mess, try simplifying the expression before you plug in.

Here's a sample problem you would do this way:

Question: Evaluate the expression $6(2x + 1) + 5x - 3(2x - 7) + 14 - 5x$ for $x=5/6$.

You could do this one by plugging in 5/6 everywhere you see an x in this expression and then doing all of the multiplications, additions, and subtractions. This, however, would take a long time and leave you prone to make a mistake when you start doing calculations like $16 + 25/6$. What's faster is to first simplify the expression to $6x + 41$ and then plug in.

They have also asked questions for which you want to use both this trick and the one from the previous section. For example,

Question: Evaluate $237\,(2x - 3y) + 3x + 5y - 235\,(2x - 3y)$ for $x=2/7$ and $y=3/2$.

The use-both-tricks approach to this one is:

- If plugging in will be a mess, try simplifying the expression before you plug in. If simplifying looks like it will also be a mess, look for terms you can group together to make it easier.

In this problem, the obvious way to group terms is $(237 - 235)(2x - 3y) + 3x + 5y$. This simplifies to $7x - y$. Plugging in is now easy and gives $2 - 3/2 = 1/2$.

Here are a couple of real IMLEM questions for practice:

Question: (IMLEM meet #1, October 2002) Evaluate the following expression for x =3/4 and y = 5/6:   12 (2x + y) –(2x – y).

Question: (IMLEM meet #1, October 2000) Evaluate the expression if x =2/7.
$$5 (13x – 8) + 2 (13x – 8) - 4 (13x – 8) + 3x + 32.$$

5.3 Solving equations in one unknown

A simple solving equations problem is:

Question: Solve for x: 3x – 6 = 3.

They won't ask a question this easy, but they might ask more some more calculation-intensive problem like:

Question: Solve for x: (5(3x + 2) – x + 3)/5 + 4 = 2(x+7) +3x – 14.

The problems of this variety don't seem to have any tricks to them. You just need to simplify the messy expression and then solve for *x*. All the questions I've seen simplify down to a simple linear equation.

One other type of question they've asked a couple times in this category are questions where you have several unknowns. For example:

Question: (IMLEM meet #1, October 2003) If 37 = 4x – 3 + 4(3x+2), y = 24 – 5x, and z = 4y – 8, find the value of z.

These problems do have something of a trick:

- Don't try to substitute in for the variables in equations like this. Also don't treat it as a system of equations: the meet 1 category description is solving single equations in a single unknown. All meet 1 problems like this can be done much more easily by just solving first for whatever variable you can find, then solving for another one, then solving for another one, and so on until you finish.

For example, in the problem above the one variable you can solve for right away is *x*. The first equation simplifies to *32 = 16x*, so *x*=2. The second equation then gives *y = 24 – 5×2 = 14*. The third then *gives z = 4×14 – 8 = 48*.

5.4 Identities

An identity is an expression that is true for all values of the variable. For example, *2 (3x + 4) = 6x + 8* is an identity.

IMLEM has asked a question involving identities for each of the last six years, so it's a good bet that they'll do so again this year. The way they usually make up an identity question is to ask you something like:

Question: Find the value of K so that the expression below is an identity:
$$2 (3x + 4) = 6x + K.$$

The answer to this is *K=8*. On a real IMLEM meet the questions will be more complicated. In each case, however, they can be done more quickly if you keep a simple trick in mind. The trick is:

- To find the value of a constant that makes an expression an identity you typically don't have to simplify the whole expression. You only need to keep track of some of the terms.

Here are a couple examples:

Question: Find the value of K so that the expression below is an identity:
2 (336x – 1) – 33 (10x) + 554x – 3 = 896x + K.

Note that in this question there is no need to ever multiply *336x* by 2, subtract *330x,* and then add *554x.* The question implies that there is some value of *K* that makes this an identity. For there to be any value of *K* that makes it an identity it must be that the left side simplifies to *896x* + something. Verifying that the terms with *x*'s add up to *896x* is a waste of time. All you need to do to answer the question is to look at the terms on the left side **that don't involve *x*** and figure out what the other terms add up to. There are just two of them: -2 and -3. Hence, the answer is that the expression is an identity for *K* = *-5.*

A real IMLEM example that works the same way is:

Question: (IMLEM Meet #1, October 2003): Find the value of K so that the expression below is an identity:
3x – 5(2x + 7) + K = 12x + 3(8 – 7x) + 2x.

Most of the terms in this expression have *x*'s in them, and we can ignore all of them when looking for the value of *K* that makes the equation an identity: this gives *-35 + K = 24.*

In the last couple of years, they've started asking problems that you might think are slightly harder. For example,

Question: (IMLEM Meet #1, October 2005): Find the value of B that will make the equation below an identity.
17(3x – 5) – (9x – 1) = B(3x – 6).

Although this might look harder at first, one could argue that it's actually easier. The key insight is that an expression like $14x - 28 = Bx - 2B$ will only be an identity only if $14=B$ **and** $28 = 2B$. What this means for you as a problem solver is that you have the option of ignoring all the terms with x's **or** ignoring all the terms without x's. Either one will give you the answer.

For example, in the IMLEM problem above you could ignore the terms with x's and get the answer by solving *17 ×(-5) + 1 = -6B*. Or you could ignore the terms without x's and finding the value of *B* that makes *51x – 9x = 3Bx* an identity. Both calculations give 14 as the answer. If presented with a problem like this I'd ignore whichever set of terms seemed more complicated. If you have time at the end of the round, you should go back and ignore the other set and make sure you get the same answer.

5.5 Made-up operations

One final type of question that they've included in this category from time to time is questions asking you to evaluate some made up operator. For example,

> *Question: (IMLEM Meet #1, October 2003) If A♣B means ($A^2 – 5B$), find the value of 11♣(7♣9).*

The answer to this question is 101. The way you get this is to first evaluate *7♣9 = 49 – 45 = 4*, and then compute *11♣4 = 121 - 20 = 101*.

Another example is

> *Question: (IMLEM Meet #1, October 2000) If A☺B means ($3A^2 – B^3$)/20. Find the value of 5☺(7☺3). Express your answer as a mixed number in simplest form.*

The answer to this question is $-7\dfrac{1}{20}$.

The way you get it is to first evaluate the expression in parentheses, *7☺3 = (147 – 27) / 20=6*, and then to compute *5☺6=(75 – 216) / 20 = -141 / 20*.

These problems don't seem to have any tricks to them. You just do the calculations they ask you to do in the order that the parentheses say to do them. One tip is that the problems have all been set up so that the first expression you evaluate turns out to be a small integer. If you get a big number or a fraction that will be very hard to plug into the next expression you should probably double check to be sure you haven't made an arithmetic error.

One final thing to note about the question is that it asked for the answer as a *mixed number*. This means a number like 1¾ rather than a number like 7/4.

IMLEM Meet #2

Congratulations on getting to meet #2! I assume from the fact that you're reading this that meet #1 couldn't have gone too badly. Either that or maybe it did go badly and that made you decide to read this section after not having read the previous one. In that case, welcome and I hope you enjoy the book.

Other than this, I don't really have anything in particular to say about meet #2. I figured I should have some text here, however, so that the format of this section would look like the others.

Category 2 – Geometry

The geometry problems in meet #2 are about areas and perimeters of polygons. There are lots and lots of facts about areas and perimeters that one could ask about on a math meet. Somehow, though IMLEM seems to ask about the same four things over and over again. If you know the four things, then geometry has often been the easiest category in meet #2. For some problems it also helps to know a little algebra.

2.1 Perimeter

The *perimeter* of a polygon is the sum of the lengths of each of its sides. For example, the perimeter of a rectangle with length ℓ and width w is $2\ell + 2w$.

- One thing you need to remember to is the Pythagorean Theorem: the sides of a right triangle satisfy $a^2 + b^2 = c^2$.

This can come in handy when finding perimeters of triangles, trapezoids, and polygons with vertices on a grid. For example, you'd need to use it over and over and over again if you wanted to find the perimeter of the figure below:

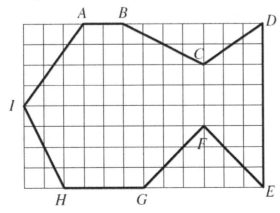

One other fact that's come up twice is that the longest diagonals in a regular hexagon are exactly twice as long as the length of the side. To see this, just draw in the lines dividing the hexagon into six equilateral triangles.

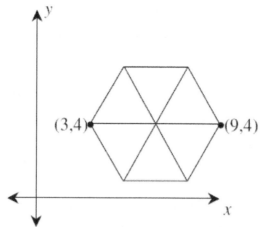

The way this has come up in questions is that they'll give you the length of a diagonal (or have you figure it out) and ask you to find the perimeter of the hexagon. In the figure above, for example, the diagonal is 6 units long, so each side of the hexagon is 3 units long, and its perimeter is 18.

2.2 Areas

The *area* of a polygon is the number of square units of space enclosed by the polygon.

The area formulas that seem to come in handy on meet #2 are

- The area of a triangle with base b and height h is A = $\frac{1}{2} b h$.

- The area of a square with side s is A = s^2.

- The area of a rectangle with length ℓ and width w is A = ℓw.

- The area of a trapezoid with bases a and b and height h is A = $\frac{1}{2} (a + b) h$.

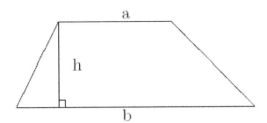

- The area of a parallelogram with base *b* and height *h* is A = *b h*.

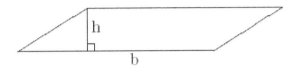

  Note that this is a special case of the trapezoid formula because a parallelogram is just a trapezoid in which the top and bottom bases are the same length. One thing to be careful about is that you need to be sure to use the height of the parallelogram and not the length of the diagonal side that connects the top and bottom.

- The area of a regular hexagon with side *s* is $A = s^2 3\sqrt{3}/2$.

- The area of a regular octagon with side s is $A = s^2(2 + 2\sqrt{2})$.

  One thing to be careful about here is that some octagons you might think are regular are not. For example the octagon shown below is not regular. The diagonal sides are $\sqrt{2}$ times as long as the vertical and horizontal sides.

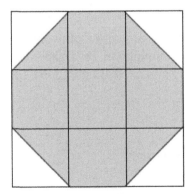

All of these formulas have come up in past IMLEM problems. If you only have time to memorize one formula I'd definitely recommend the trapezoid formula – the rectangle, parallelogram, and triangle formulas are all special cases (think of a triangle with the top base having length zero). In addition the trapezoid comes up over and over and over again. For example, in 2000 they asked:

Question (IMLEM Meet #2, December 2000) The pattern pictured below is to be cut form fabric that comes on a bolt that is 42 inches wide. If Sue buys 2 yards of fabric from the 42 inch bolt and cuts out her pattern, how many square inches of fabric will be wasted?

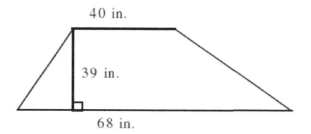

And in 2002 they asked:

Question (IMLEM Meet #2, November 2002) If the trapezoid shown here has an area of 195.5 square centimeters. How many centimeters are in the length of base one.

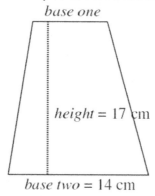

The answer to the first problem is $42 \times 72 - 39 \times (40 + 68)/2 = 918$. (The arithmetic is easier if you realize that both products are multiples of 18. To answer the second note that $195.5 = \frac{1}{2}$ (base one + 14) 17 implies that $391 / 17 =$ (base one + 14). Using $391 = 17 \times 23$, we find that the answer is 9.

2.3 Perimeters of Rectilinear Figures

A rectilinear figure is a polygon in which all angles are right angles. One finds lots of them on IMLEM meets. For example, the 2000 meet included this shape:

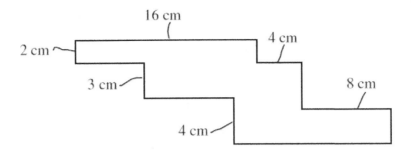

And the 2003 meet actually had two such shapes, one of which is shown below.

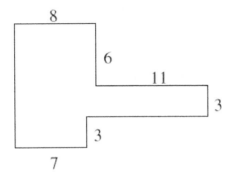

The questions about these figures always boil down to the same thing: what is the perimeter? Answering these questions is easy once you've seen the trick.

2.3.1 City block distance

Suppose that you live in a city with a square grid system of streets. Suppose that you start out at the black dot on at the lower left corner of the map shown below and want to walk to the point in the upper right hand corner. There are lots of ways to do this. For example, you could first walk six blocks north and then walk six blocks east. Or, you could first walk five blocks east, then turn north for two blocks, walk one more block to the east, then four more blocks north. Or you could go one block north, then one block east, then one block north, etc. (To be clear the straight line in the figure is not allowed. We assume you can't walk through buildings.)

What is the shortest path?

The answer is that there are lots of shortest paths. Any path that doesn't involve any backtracking will have the same length: $\ell = n + e$, where n is the number of blocks north you need to go and e is the number of blocks east you need to go.

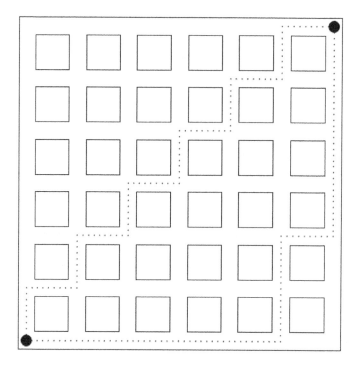

2.3.2 Around-the-city walks

Suppose that the grid above is the complete city map and you want to go on a walk reaches the northernmost, westernmost, southernmost, and easternmost extremes in the city and finishes at your starting point.

Again, there are many ways to do this. You could start in one of the corners and trace out a rectangle by following the city's perimeter. Or you could start in the lower left corner, use the north-east-north-east-etc. path to reach the upper right corner, then head back along the path that involves going south for four blocks, west for one, south for two, and west for five.

Which path is the shortest?

Again, the answer is that there are many shortest paths. Any path that involves no backtracking will have length $\ell = 2(n + e)$, where n is the north-south size of the city and e is the east-west size of the city. What does "no backtracking" means? Basically it means that if you start from the northernmost point on the path the path first heads exclusively south and west, then (after you've reached the western boundary) heads exclusively south and east, then (after you've reached the southern boundary) heads exclusively north and east, then exclusively north and west.

2.3.3 Questions about perimeters

What does all of this have to do with IMLEM questions?

The answer is that IMLEM questions on rectilinear figures always ask you to find the perimeter, and that most of the time the boundaries of the figures can be though of as around-the-city walks that are as short as possible.

For example, the first figure in this section traces out an around-the-city walk in a city that has a north-south distance of 9cm and an east-west distance of 28 centimeters. It involves no backtracking, so its perimeter is 2(9 + 28) = 74 cm.

The second figure (from 2003) similarly traces out an around-the-city walk. This time, the city is 12 units in the north-south direction and 19 units from east to west. The perimeter is 2 (12 + 19) = 62.

Most of the time all that is involved in doing these problems is to look carefully and add up distances to find the north-south and east-west distances.

I should mention, however, that one time they tried to trick students by including a figure that does involve backtracking.

Question: Find the perimeter of the figure below.

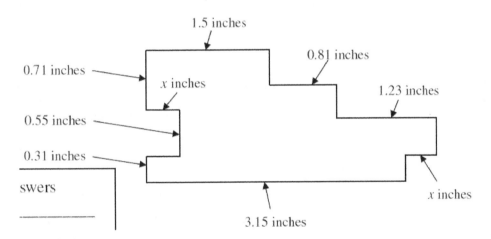

Do you see where the backtracking is in the figure below? The backtracking is the little jog that goes x units to the east and then x units back to the west while traversing the western boundary. Hence, the perimeter is $2x$ units longer than the shortest around-the-city walk would have been. The shortest around the city walk would be 2 (1.57 + 3.54) = 10.22 inches, so the perimeter is 10.22 + 2x. To finish the problem, you also need to figure out what x is. The way you do this is to notice that the east-west distance along the top side is 1.5 + 0.81 + 1.23 = 3.54 inches, whereas the one on the bottom is 3.15 + x inches. This gives $x = 0.39$ so the perimeter is 10.22 + 2x = 11 inches. It's a good feeling when you get an answer this simple after adding so many complicated decimals. It's a pretty good indicator that you got it right.

2.4 Problem Solving Tip: Subtract, Subtract, Subtract

The final thing that they really like to ask is problems in which you have to find the area of a shaded region. In almost every case the easiest way to do this is by subtraction. You first find the area of the big rectangle they started with, and then subtract off the area of the figures that are not shaded.

Here are two examples:

Question (IMLEM Meet #2, November 2001). In the rectangle below, the measure of AB is 24 inches, the measure of AD is 11 inches, and the measure of EF is 10 inches. If ED is parallel to FG, how many square inches are in the combined area of the shaded regions?

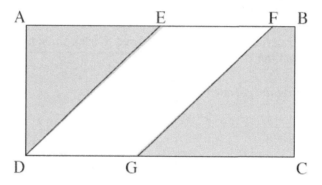

The answer to this problem is 154. The area of the rectangle is $11 \times 24 = 264$. The parallelogram DEFG has a base of 10 (the length of EF) and a height of 11 (the length of AD). Hence, its area is 110 square inches and the area of the shaded region is $264 - 110 = 154$ square inches.

Another way to do this problem would have been to slide the two gray parts together in your head. If you're good at doing this you'd figure out that sliding them together gives a rectangle that's 14 inches across the top and 11 inches high. This gives a quicker path to the answer: $11 \times 14 = 154$. Be careful with doing things like this though. People who make up math contest questions are devious. They can be good at coming up with questions where you think it's obvious what two shapes will look like after they are spun around or slid together, but where the obvious answer is wrong.

Question (IMLEM Meet #2, November 2004) Three right triangles and two squares were cut out of a rectangle with length 20 centimeters and width 12 centimeters as shown below. The dimensions of the triangles are given in the picture and the squares are each 3.5cm by 3.5cm. How many square centimeters are in the area of the shaded region?

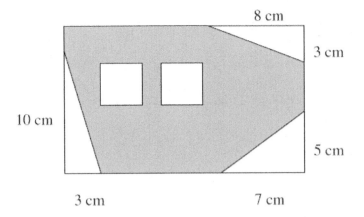

Doing the arithmetic for this one is a bit of a pain, but at least by telling you the triangles and squares were cut out they made it more obvious that you need to do some subtraction. The rectangle has area 20 × 12 = 240. The areas of the three triangles are ½ (3 × 10) = 15, ½ (8 × 3) = 12, and ½ (5 × 7) = 17.5. Each square has area (7/2) ×(7/2) = 12.25.

The answer is 240 – (15 + 12 + 17.5 + 2×12.25) = 171.

2.5 Advanced Topic: More Area Formulas

There are lots of other neat area formulas that would let you easily solve some problems that you otherwise would have no chance of solving. IMLEM hasn't asked about them yet, but we can hope that this year will be different.

2.5.1 Triangles

There are lots of formulas for finding the area of a triangle. Four nice ones are:

- Heron's fomula: $A = \sqrt{s(s-a)(s-b)(s-c)}$ where a, b, c are the lengths of the three sides and s is the semiperimeter, $s = (a + b + c)/2$.

The great thing about this formula is that all you need to know to apply it is the lengths of the three sides. You don't need to know the height of the triangle.

An example of a problem where this really helps is:

Question: Find the area of a triangle if the lengths of its sides are 4, 5 and 7.

The semiperimeter is (4 + 5 + 7)/2 = 8, so Heron's formula gives the answer as $\sqrt{8 \times 4 \times 3 \times 1} = 4\sqrt{6}$.

And example of a problem where Heron's formula might seem like a good one to use, but where it's not the easiest way to get the answer is:

Question: Find the area of a triangle if the lengths of its sides are 8, 15 and 17.

- The area of an equilateral triangle with side s is $A = s^2\sqrt{3}/4$.

- $A = r\,s$, where r is the radius of the inscribed circle and s is the semiperimeter.

- $A = \dfrac{abc}{4R}$, where R is the radius of the circumscribed circle.

2.5.2 Regular pentagon

The area of a regular pentagon with side s is $\dfrac{1}{4}s^2\sqrt{25+10\sqrt{5}}$. I can't imagine that they'd ask about this, but I like the fact that it has a square root within a square root.

2.5.3 Regular n-gon

A formula that works for any regular n-gon is $A = \dfrac{1}{4}ns\sqrt{4R^2 - s^2}$, where R is again the radius of the circumscribed circle. Two special cases where you might be able to check that this formula works are the square and the hexagon.

I discuss a couple other very neat results in the section titled "Areas of Polygons on a Grid" in the MathCounts part of this book. They're probably unlikely to come up in the geometry category of Meet #2, but could appear on a team round. It'd be good if at least one person on your team knew them.

Category 3 – Number Theory

The number theory category in meet #2 covers prime factorization, greatest common factors and least common multiples. It's been an easy category to get some points on, but a hard category to get a 6 on: there are usually one or two questions that are pretty straightforward (if you know the material in these notes), but there has also usually been one harder question that I can't tell you how to do.

The prime factorization part has included questions on most of the topics from meet #1: prime and composite numbers; prime factorization; factor tables; divisibility rules, numbers of factors; sums of factors; etc. I'll give a quick review of some of these at the end of this section, but you should go back to the meet #1 packet if you don't remember them well.

3.1 A Super-Quick Review of Prime Factorization

Recall that any number can be uniquely factored in to a product of primes. For example, $594 = 2 \times 3 \times 3 \times 3 \times 11$, $2006 = 2 \times 17 \times 59$, $2007 = 3 \times 3 \times 269$, and $2008 = 2 \times 2 \times 2 \times 251$.

A convenient way to write prime factorizations is in exponential form. For example, we can write the factorization of 594 as $594 = 2^1 3^3 5^0 7^0 11^1$.

The *factors* of 594 are all of the positive integers that can be multiplied by another integer to get 594. For example 11 is a factor because $11 \times 54 = 594$. One can find all of the factors of a number by making factor tables using the prime factorization. For example, the factor tables for 594 are:

1	1	3	9	27
1	1	3	9	27
2	2	6	18	54

11	1	3	9	27
1	11	33	99	297
2	22	66	198	594

3.2 Greatest Common Factors

The *greatest common factor* of a and b, written GCF(a,b), is the largest number that is a factor of a and a factor of b. For example, the factors of 12 are 1, 2, 3, 4, 6, and 12, and the factors of 30 are 1, 2, 3, 5, 6, 10, 15, and 30, so GCF(12, 30) =6.

The best way to find common factors is rarely to compute all of the factors of the two numbers. Instead, you should try to use one of two methods.

3.2.1 Prime factorization method

65

The first good method for finding greatest common factors is to find the prime factorizations of the two numbers and use the following result:

Proposition: Suppose that $n = 2^{a_n} 3^{b_n} 5^{c_n} 7^{d_n} \ldots$ and $m = 2^{a_m} 3^{b_m} 5^{c_m} 7^{d_m} \ldots$ Then, $\text{GCF}(n,m) = 2^{\min\{a_n,\, a_m\}} 3^{\min\{b_n,\, b_m\}} 5^{\min\{c_n,\, c_m\}} 7^{\min\{d_n,\, d_m\}} \ldots$

Here are a couple examples of how this works:

Question: Find the greatest common factor of 90 and 594.

Question: Find the greatest common factor of 102 and 2006.

Start by factoring both numbers in each pair. In the first problem we have $90 = 2^1 3^2 5^1$ and $594 = 2^1 3^3 5^0 7^0 11^1$. Hence, $\text{GCF}(90,594) = 2^1 3^2 = 18$. In the second problem we have $102 = 2^1 3^1 17^1$ and $2006 = 2^1 17^1 59^1$, so $\text{GCF}(102,2006) = 2^1 17^1 = 34$.

3.2.2 Euclidean division

When the numbers in the problem are big and unfamiliar it can take a long time to find the prime factorization. In such cases I find the following reduction method much easier. This method relies on applying the following result over and over.

Proposition: If n is a multiple of m, then $\text{GCF}(n,m) = m$. Otherwise, $\text{GCF}(n,m) = \text{GCF}(m, n - km)$ for any k.

The formal statement can be a little confusing, but the method is really very easy to learn and use. Here are some examples:

Question: Find the greatest common factor of 2008 and 2006.
Question: Find the greatest common factor of 126 and 78.
Question: Find the greatest common factor of 1400 and 135.

Applying the method to the first problem just involves writing

$\text{GCF}(2008,2006) = \text{GCF}(2006, 2008 - 2006) = \text{GCF}(2006,2) = 2.$

In the second problem you need to repeat a calculation like this a few more times:

$\text{GCF}(126,78) = \text{GCF}(78,126 - 78) = \text{GCF}(78,48) = \text{GCF}(48,30) = \text{GCF}(30,18) = \text{GCF}(18,12) = \text{GCF}(12,6) = 6.$

Of course, you could have stopped as soon as you recognized that the answer was 6.

The third problem is an example where subtracting 135 from 1400 repeatedly would take a long time. To make the numbers get small more quickly, what you want to do here is to subtract a multiple of 135 from 1400. The obvious multiple to use is $10 \times 135 = 1350$. This gives:

$GCF(1400, 135) = GCF(135, 1400 - (10 \times 135)) = GCF(135, 50) = GCF(50, 35) = 5.$

For practice you could go back and redo the examples I did via the prime factorization method. Obviously, the answer should be the same either way.

This method is sometimes called Euclidean division and is used in other parts of math too.

3.3 Least Common Multiples

The *least common multiple* of a and b, written LCM(a,b), is the smallest number that is a multiple of a and a multiple of b. For example, LCM(12, 30) = 60.

The best way to find least common multiples again often starts with the prime factorization:

Proposition: If $n = 2^{a_n} 3^{b_n} 5^{c_n} 7^{d_n} \dots$ and $m = 2^{a_m} 3^{b_m} 5^{c_m} 7^{d_m} \dots$, then LCM($n,m$) = $2^{max\{a_n, a_m\}} 3^{max\{b_n, b_m\}} 5^{max\{c_n, c_m\}} 7^{max\{d_n, d_m\}} \dots$

Here are a couple examples of how this works:

Question: Find the least common multiple of 90 and 594.

Question: Find the least common multiple of 102 and 2006.

In the first problem we have $90 = 2^1 3^2 5^1$ and $594 = 2^1 3^3 5^0 7^0 11^1$. Hence, LCM(90,594) = $2^1 3^3 5^1 11^1 = 2970$. In the second problem we have $102 = 2^1 3^1 17^1$ and $2006 = 2^1 17^1 59^1$. Hence, LCM(102,2006) = $2^1 3^1 17^1 59^1 = 6018$.

3.4 More on GCFs and LCMs

There are a few other facts about GCF's and LCM's that seem to be fairly useful to know.

3.4.1 GCF-LCM product formula

The GCF-LCM product formula has been the single most useful thing to know for this category. At one point is seemed like they had a question on it every single year. These days, they don't always have it on, but it must still be a good bet. The theorem is

Proposition: GCF(n,m) $\times$ LCM(n,m) = $n \times m$.

A typical example of a question where this comes in extremely handy is

Question (IMLEM Meet #2 variant, December 2001): Suppose that GCF(x,y)=5 and LCM(x,y)=225. If x=25, what is y?

The theorem tells us that $(225 \times 5) = 25y$, so the answer is $(225 \times 5)/25 = 45$.

On the actual meet they had LCM(x,y)=525 and x=75. The answer to that problem is $(525 \times 5)/75 = 35$. Here, it helps to be lazy and look for things to cancel instead of multiplying and then doing long division. Both 525 and 75 are obviously multiples of 25 so you'd start there. I changed the problem from the original one semi-accidentally. When I first copied down the problem I had x=75, but accidentally wrote 225 instead of 525. It was only later that I noticed that my answer didn't work. The GCF-LCM formula says the answer must be $(225 \times 5)/75 = 15$, but it's not: LCM(15, 75)$\neq$225. The method doesn't work because there is no answer: there is no number y with GCF(75, y)=5 and LCM(75,y)=225. In the first printing I suggested that Kent missed this when he made up the problem. But he didn't. It was just my mistake. A good lesson here is that if you think a problem doesn't work, then go back and check to make sure you didn't misread something or copy some number down wrong.

   Question: Let a=GCF(42,102) and b=LCM(42,102). What is a×b?

This is another problem where the GCF-LCM formula makes things much quicker. The brute force approach would be to first find the GCF, then find the LCM, then multiply. The GCF-LCM product formula lets you skip the first two steps. The answer is $42 \times 102 = 4284$.

3.4.2 GCF's and LCM's of larger sets

The prime factorization method for finding GCF's and LCM's extends naturally to a method for finding the GCF and LCM of a whole set of numbers. You just take the minimum or maximum of all of the exponents. For example,

   Question (IMLEM Meet #2, December 2000): Find the GCF and LCM of the set {12, 36, 11, 18, 24}

The prime factorizations of these numbers are:

$12 = 2^2 3^1$ $18 = 2^1 3^2$
$36 = 2^2 3^2$ $24 = 2^3 3^1$
$11 = 11^1$

Using this we find GCF(12,36,11,18,24) = 1 and LCM(12,36,11,18,24) = $2^3 3^2 11^1 = 792$.

Obviously taking the time to factor all the numbers wasn't really necessary to find the GCF. You should have realized that the answer was 1 as soon as you saw that they included 11 (a prime number that is not a factor of all other numbers) in the list.

Another way to find the LCM that might have been easier to think about is to use the recursive formula:

* LCM(a,b,c) = LCM(a,LCM(b,c)).

In a five number problem, there are lots of variants that work. For example,

LCM(a,b,c,d,e) = LCM(a,LCM(b,LCM(c,LCM(d,e)))) and
LCM(a,b,c,d,e) = LCM(a,LCM(b,c), LCM(d,e)))

I know you're all thinking, "How in the world could a formula like that be useful or make something easy?" You may be right. If you're not though here's the reason: when you apply this formula you can choose in which order to put the numbers and some orderings make the problem easier than others. In the problem above, the order I'd put them in is 11, 24, 12, 18, 36. The answer is then

LCM(11, 24, 12, 18, 36) = LCM(11,LCM(24,12),LCM(18,36))
                         = LCM(11,24,36))
                         = LCM(11,LCM(24,36))
                         = LCM(11,72)
                         = 11 × 72 = 792.

3.4.3 Factor counting problems in disguise

Back on meet #1 we learned how to count the number of factors of a number and to do some variants of this (like counting how many even factors of a number). Some problems involving LCMs are just problems like this in disguise. For example,

Question: For how many values of n is LCM(n,54) = 54.

This is a standard factor counting problem. LCM(n,54) = 54 only if n is a factor of 54. Because $54 = 2^1 \, 3^3$, the number of factors of 54 is $(1+1)(3+1) = 8$.

An example of a slightly harder problem would be:

Question: For how many values of n is LCM(n,54) = 540.

Note that $54 = 2^1 \, 3^3$ and $540 = 2^2 \, 3^3 \, 5^1$. Hence, we know that n must be of the form $2^2 \, 3^x \, 5^1$ with $x \leq 3$. There are four possibilities corresponding to x=0, 1, 2, and 3.

3.4.4 Other word problems

A couple of different kinds of word problems have come up in the number theory category of meet #2. One of them is problems involving cycles of different lengths. For example,

Question: Alice, Bob, and Charlie are running laps at on a track. They start from the same point on the track at 12 noon and all run counterclockwise. Alice runs one lap of the track every 6 minutes. Bob runs one lap every 8 minutes. Charlie runs one lap every 10 minutes. What the next time at which they will all simultaneously arrive at the starting point?

Alice arrives back at the starting point at every multiple of six. Bob arrives back at every multiple of 8. Charlie arrives back at every multiple of ten. Hence, the number of minutes until all three arrive back simultaneously is LCM(6,8,10) = 120. This makes the answer 2pm (at which point Alice, Bob, and Charlie will be very tired!)

A second type of problem is a square/cube filling a poster/box problem. For these you need to compute GCFs. An example is

Question: The interior of a cardboard box measures 72cm long, 54cm wide, and 42cm deep. What are the dimensions of the largest cube that could be used to fill up the box leaving no extra room?

A cube with integer-length sides will fill the box exactly if and only if the side length is a common factor of 42, 54, and 72. The answer is GCF(42, 54, 72) = 6. They've also had similar questions asking about sizes of squares that would exactly cover a poster. Next time they might ask about tiles that could be used to cover a bathroom floor without cutting any tile.

3.4.5 Largest and smallest factors/ products

A few questions have also asked about what factors would make some expression as large or as small as possible. The simplest of these are questions like:

Question: A proper factor is a factor that is different from the number itself. What is the largest proper factor of 240?

The answer is 120. You get it by dividing 240 by its smallest factor, which is 2. A harder problem they've asked sometime is

Question: What is the smallest possible value for a+b if LCM(a,b)=420?

Note that $420 = 2^2 \, 3^1 \, 5^1 \, 7^1$, so either a or b must be a multiple of 2^2, one must be a multiple of 3, etc. If we're trying to make $a + b$ as small as possible, then obviously we don't want to repeat any of these factors in both a and b. Hence, the only remaining question is how to divide up the prime factors between a and b. A rule-of-thumb for doing this is that you want to make each number about the same size. Here, the best way to do this is to divide them into $2^2 \times 5 = 20$ and $3 \times 7 = 21$. The answer is indeed $20 + 21 = 41$.

3.4.6 LCM's with noninteger arguments

Sometimes you'll see variants of LCM problems where the numbers are not integers. For example,

Question: Alice and Bob are running laps at on a track. They start from the same point on the track at 12 noon and all run counterclockwise. Alice runs one lap of the

track every 1⅓ minutes. Bob runs one lap every 2½ minutes. What the next time at which they will all simultaneously arrive at the starting point?

One way to do this is to just convert to seconds. Alice's laps take 80 seconds. Bob's take 150 seconds. You can compute LCM(80, 150) and then convert back to minutes.

The fact that minutes come in 60 seconds is obviously not necessary for this technique to work. For example, if Alice took 3/7 of a minute and Bob took 2/9 of a minute neither one would be a whole number of seconds. What you could do, however, is to define your own measure a short-second equal to one-sixty-third of a minute. Alice would then do a lap every 27 short-seconds and Bob a lap every 14 short-seconds. The answer is LCM(27, 14) = 27 × 14 short-seconds, which is equal to 27 × 14 / 63 = 6 minutes.

You can save time by not bothering to give a name to the units you divide a minute into.

3.5 A Slightly Longer Review of Prime Factorization

As I noted above, the first thing to remember is that every number can be uniquely factored as a product of primes and these prime factorizations can be written compactly using exponential notation. For example,

$$44 = 2 \times 2 \times 11 = 2^2\ 11$$
$$540 = 2 \times 2 \times 3 \times 3 \times 3 \times 5 = 2^2\ 3^3\ 5$$

The easiest way to find prime factorizations is usually just to keep looking for one factor at a time and adding them to what you've found. For example,

$$\begin{aligned} 84 &= 2 \times 42 \\ &= 2 \times 2 \times 21 \\ &= 2 \times 2 \times 3 \times 7 \end{aligned}$$

When you're doing this it helps to use rules you know for divisibility.

- A number is a multiple of 2 if the last digit is 0, 2, 4, 6, or 8.
- A number is a multiple of 5 if the last digit is 0 or 5.
- A number is a multiple of 10 if it ends in a zero.

- A number is a multiple of 3 if the sum of the digits is a multiple of 3.
- A number is a multiple of 9 if the sum of the digits is a multiple of 9.

- The number $abcde$ is a multiple of 2 if e is multiple of 2.
- The number $abcde$ is a multiple of 4 if $2d + e$ is multiple of 4.
- The number $abcde$ is a multiple of 8 if $4c + 2d + e$ is multiple of 8.

- The number $abcde$ is a multiple of 11 if the alternating sum/difference of the digits is a multiple of 11, i.e. if $e - d + c - b + a$ is a multiple of 11. For example, the number 14641 is a multiple of 11: $1 + 4 - 6 + 4 - 1 = 0$

If you want to find all of the factors of a number the best thing to do is to make a factor table. For example, the factor table of 200 is:

	1	2	$2^2=4$	$2^3=8$
1 |	1	2	4	8
5 |	5	10	20	40
$5^2=25$ |	25	50	100	200

When a number has more than two distinct prime factors you need to make multiple tables. For example, $4200 = 2^3 \cdot 3 \cdot 5^2 \cdot 7$ has 48 factors:

1 |	1	2	4	8
1 |	1	2	4	8
5 |	5	10	20	40
25 |	25	50	100	200

3 |	1	2	4	8
1 |	3	6	12	24
5 |	15	30	60	120
25 |	75	150	300	600

7 |	1	2	4	8
1 |	7	14	28	56
5 |	35	70	140	280
25 |	175	350	700	1400

3×7 |	1	2	4	8
1 |	21	42	84	168
5 |	105	210	420	840
25 |	525	1050	2100	4200

If you only want to count how many factors a number has, then you don't need to write out the table. You can either think about what it will look like and multiply (number of rows) × (number of columns) × (number of tables) or use the following formula:

- If the prime factorization of n is $n = 2^a \cdot 3^b \cdot 5^c \cdot 7^d \cdot 11^e \cdot 13^f \cdots$, then the number of factors of n is $(a+1)(b+1)(c+1)(d+1)(e+1)(f+1)\ldots$

This formula seems to come in handy pretty often.

The formula for finding sums of factors is:

- If the prime factorization of n is $n = 2^a \cdot 3^b \cdot 5^c \cdot 7^d \cdot 11^e \cdot 13^f \cdots$, then the sum of the factors of n is $(1 + 2 + 2^2 + \ldots + 2^a)(1 + 3 + 3^2 + \ldots + 3^b)(1 + 5 + \ldots) \ldots$

For example, the prime factorization of 200 is $n = 2^3 \cdot 5^2$, so the sum of its factors is $(1 + 2 + 4 + 8)(1 + 5 + 25) = 15 \times 31 = 465$.

Category 4 – Arithmetic

The arithmetic category in meet #2 covers fractions, percents, and terminating and repeating decimals. The fractions and percents part is not very hard. The terminating and repeating decimals part covers a lot of material not normally taught at Bigelow.

4.1 Fractions and Percents

All of the past IMLEM meets have included at least one fractions and percents problem. These problems usually just ask you to do a series of multiplications and divisions. They're not hard *if* you understand the questions and are pretty good at multiplying and factoring.

4.1.1 "Of" problems

One thing they seem to like to do is to ask word problems using the word "of." For example,

Question: What number is 2/3 of 45?
Question: What is 20% of 80?
Question: What is 40% of 2/3 of 0.7 of $900?

The main thing you need to know to do these problems is that "of" means "times." For example, "2/3 of 45" means $2/3 \times = 45 = 30$. In the second question is that 20% of 80 means 0.20×80 which is 16. The answer to the last question is $0.4 \times 2/3 \times 0.7 \times \$900 = \$168$.

4.1.2 "Greater than" problems

Another popular form of question is word problems involving the phrase "greater than."

Question: What number is 2/3 greater than 45?
Question: What fraction is 40% greater than 2/9?

What "greater than" means in a problem like this is to multiply the two numbers together and add the result to the second number. For example, the number that is "2/3 greater than 45" is $(2/3 \times 45) + 45 = 75$. An equivalent formula for this is to add one to the first number and then multiply, e.g. number that is "2/3 greater than 45" is $(1 + 2/3) \times 45 = (5/3) \times 45 = 75$. Problems involving fractions are usually easier to do this way. For example, the answer to the second question is $1.4 \times 2/9 = 7/5 \times 2/9 = 14/45$.

4.1.3 Dividing fractions

They also ask some problems about dividing fractions. The way to do this is to invert and multiply, e.g. $(5/6) \div (5/9) = (5/6) \times (9/5) = 3/2$.

4.1.4 Three tips

73

Some of the problems in this category are designed to take a long time of you try to do all the multiplications and to go faster if you think of a slight trick. One example is:

Question: What the difference between 37.5% of 72 and ¼% of 6400?

The trick in this problem is to us fractions instead of decimals. 37.5% is 3/8. ¼% is 1/400. Hence the answer is $(3/8 \times 72) - (6400/400) = 27 - 16 = 11$.

A second example with a different trick is:

Question: What is 35% of 6-and-two-thirds times 3/55 of 121.

The trick to this problem is to write each term as a fraction and look for terms to cancel before doing any multiplication. Here, *35% of 6-and-two-thirds times 3/77 of 121* means $\frac{7}{20} \times \frac{20}{3} \times \frac{3}{77} \times 121$. Cancelling the 20's and 3's and factoring 121 this is $\frac{7}{77} \times 11 \times 11$. You can then cancel a 7 from the top and bottom of the first fraction and a pair of 11's to get that the answer is 11.

A third tip is that you should be sure to reduce all fractions. Many problems are designed so that if you do the calculations the most obvious way you'll end up with an answer like 34/119. This answer would be marked wrong regardless of what the question is. Why? Well, because it's not reduced: the numerator and denominator are both multiples of 17.

4.2 Terminating and Repeating Decimals

IMLEM meet #2's have also all included at least one problem on repeating decimals. The first thing you need to know is what a repeating decimal is.

4.2.1 Definitions

A *terminating decimal* is a standard decimal number that you can write out in its entirety. For example, one-quarter can be written as 0.25 and 17.453 can be written as 17.453.

Other rational numbers can't be written this way. The most familiar example is one-third. Its decimal representation goes on forever: 0.33333333333333333333333... A decimal expression that goes on forever with a regular pattern like this is called a *repeating decimal*. Writing repeating decimals as I did above has two drawbacks: it takes up a lot of room; and you could get confused about exactly what the ... means. This has led mathematicians to invent a better notation: they put line over the part that repeats. For example they'd write one-third as $0.\overline{3}$ and 0.27373737373737373... as $0.2\overline{73}$.

There's a simple rule for figuring out which fractions terminate and which repeat.

- If all prime factors of the denominator (of a simplified fraction) are 2's and 5's, then the fraction can be written as a terminating decimal.

- Otherwise it can not. It will either be a pure repeating decimal like $0.\overline{3}$ and $0.\overline{73}$, or something like $0.2643\overline{573}$, which starts out with a few patternless digits and then repeats.

4.2.2 Some common repeating decimals

You should immediately recognize some repeating decimals. The most familiar, of course, are the thirds:

$1/3 = 0.\overline{3}$
$2/3 = 0.\overline{6}$

The sixths also come up a lot. Note that the first digits in these decimal expansions are different from all the ones that come afterwards.

$1/6 = 0.1\overline{6}$
$5/6 = 0.8\overline{3}$

A couple others you sometimes see are

$1/7 = 0.\overline{142857}$
$1/11 = 0.\overline{09}$

The one-seventh formula is neat because the digits rotate around as you change the numerator:

$1/7 = 0.\overline{142857}$
$2/7 = 0.\overline{285714}$
$3/7 = 0.\overline{428571}$
$4/7 = 0.\overline{571428}$
$5/7 = 0.\overline{714285}$
$6/7 = 0.\overline{857142}$

The most important repeating decimals, however, are the ninths, ninety-ninths, etc.

$1/9 = 0.\overline{1}$
$1/99 = 0.\overline{01}$
$1/999 = 0.\overline{001}$
$1/9999 = 0.\overline{0001}$

4.2.3 Converting repeating decimals to fractions: the 999 method

The reason why the formulas for 1/9, 1/99, etc. are so important is that you can use them to figure out what any pure repeating decimal is. For example

$0.\overline{5} = 5 \times 0.\overline{1} = 5/9.$
$0.\overline{27} = 27 \times 0.\overline{01} = 27/99 = 3/11.$
$0.\overline{047} = 47 \times 0.\overline{001} = 47/999.$

The most common IMLEM problem is some calculation that you do by converting repeating decimals to fractions. For example,

Question: Simply $0.\overline{21}/0.\overline{56}$. Write you answer as a fraction in lowest terms.

The first step in answering this is to replace the repeating decimals with 21/99 and 56/99. The second is to cancel the 99's to get 21/56. The third is to simplify this to 3/8.

What do you do if the decimal you're supposed to simplify is not a pure repeating decimal?

Question: Express $0.1\overline{63}$ as a fraction in lowest terms.

One way to extend the 999 technique is to solve problems like this is to recognize that

$0.1\overline{63} = 0.1 + 0.\overline{63} / 10 = (1/10) + (63/990) = 162/990 = (18 \times 9)/(9 \times 110) = 9/55.$

Here's one more relatively hard problem of this variety to try at home:

Question: Express $0.2\overline{037}$ as a fraction in lowest terms.

The answer is 11/54.

4.2.4 Converting repeating decimals to fractions: the algebra method

If you're comfortable with algebra, you can also do the conversions another way. Pure repeating decimals are easy with this method. For example, to convert $0.\overline{27}$ you first write

$x = 0.\overline{27}$

Then multiply by 100 to get

$100x = 27.\overline{27}$

Subtract the two gives

$99x = 27$,

Which we solve for $x = 27/99 = 3/11$.

The algebra method wasn't necessary here, but it can help you to avoid getting confused when solving more complicated problems. For example, to convert $0.16\overline{3}$ you'd write

$x = 0.16\overline{3}$
$10x = 1.6\overline{3}$
$1000x = 163.\overline{63}$

Then subtract the second line from the third to get

$990x = 162$
$x = 162/990 = 9/55$

Some people find this method much more confusing than the 999 method. If you're one of them, don't worry about it. Just get out a pen and draw a big X through this section to remind yourself not to waste time reading it again when you're reviewing for the meet.

4.2.5 Writing fractions as repeating decimals

In some sense, this is very easy. You can find the decimal expansion of any fraction just by doing long division. This is the way to do some problems. For example, try doing this on 1/37.

I mean it. You won't get used to doing this unless you try doing long division on a few fractions. I'll waste some time here so you actually do it before just looking at the answer.

Hopefully, by now you did the division and found the pattern pretty quickly: $1/37 = 0.\overline{027}$.

In other problems, however, it will take you much longer before you notice a pattern. For this reason, converting fractions to decimals is usually the harder direction to go on a math contest.

One trick that sometimes lets you do this more quickly is

- Think about whether the denominator is a factor of 9, 99, 999, 9999 or some other number.

The decimal expansion of 1/37 is probably the best example of this trick. $37 \times 3 = 111$. Hence, we know $37 \times 27 = 999$ and $1/37 = 27/999 = 0.\overline{027}$.

Another problem you could try out that works similarly is 1/27. A slightly harder example is

Question: Express 1/303 as a repeating decimal.

If you're having trouble with this one a hint is to try 9999 as the denominator.

To help yourself think quickly about problems like these it may be useful to memorize some prime factorizations:

$9 = 3 \times 3$
$99 = 3^2 \times 11$
$999 = 3^3 \times 37$
$9999 = 3^2 \times 11 \times 101$
$99999 = 3^2 \times 41 \times 271$
$999999 = 3^3 \times 7 \times 11 \times 13 \times 37$
$9999999 = 3^2 \times 239 \times 4649$
$99999999 = 3^2 \times 11 \times 73 \times 101 \times 137$

A second related trick is

- Look for relations to problems you know.

A simple example of this technique is

Question: Express 5/60 as a repeating decimal.

The way to do this is to notice that $5/60 = (1/10) \times 5/6 = (1/10) \times 0.8\overline{3} = 0.08\overline{3}$. A really hard example is

Question: Express 3/14 as a repeating decimal.

The first step is to realize that because 70ths are kind of like 7ths you might be able to solve this by converting to 70ths: 3/14 = 15/70. What makes this a really hard problem is that you then need to make a second clever observation:

$$15/70 = 14/70 + 1/70 = 1/5 + (1/10)\,1/7 = 0.2\overline{142857}.$$

(and you also need to remember the repeating decimal form for 1/7).

4.2.6 Figuring out repeat lengths

Another very popular type of question on IMLEM is to ask for a very far off digit of some fraction. For example, they might ask:

Question: What is the 50th digit to the right of the decimal point in the decimal expansion of 1/7?

Question: What is the 97th digit to the right of the decimal point in the decimal expansion of 1/17.

In these problems they usually pick a digit that is far enough along in the decimal expansion so as to make it obvious that you could not possibly do the full division even with a calculator (which is not allowed on meet #2). Instead,

- The way to approach these problems is to think about the repeat pattern.

In the case of 1/7, for example, you know that it repeats every sixth digits. Hence, the 1st digit is the same as the 7th, and the 13th, and the 19th, and so on. Because 50 is two more than a multiple of 6, the 50th digit is the same as the 2nd. The answer is 4.

A variant of this problem that's good to think about is

Question: What is the 50th digit of 1/70?

The answer is 1.

The question about the 97th digit of 1/17 is harder because most of you probably don't already know what the repeat pattern is for 1/17. A theorem that comes in very handy in problems like this is:

- *Theorem: If p is a prime number (other than 2 or 5) and a is not a multiple of p, then a/p is a pure repeating decimal and the length of the repeat is a factor of p – 1.*

In the 1/17 example, the theorem tells you that the repeat length is a factor of 16. This implies that the 1st digit in the decimal expansion of 17 is the same as the 17th, and the 33rd, and the 49th, and the 65th, and the 81st, and the 97th. The first digit in the expansion of 1/17 is zero, so the answer is 0.

One additional trick you can use is to recognize that the fact that a problem is answerable gives away what the answer must be. For example on the 2005 meet they asked:

Question (IMLEM Meet #2, Nov. 2005): What is the 53rd digit to the right of the decimal point in the decimal expansion of 9/37.

The first theorem I gave you implies that the 9/37 is a pure repeating decimal and the repeat length is a factor of 36. If the repeat length were 36, then the 53rd digit would be the same as the 17th and to do the problem you'd have to do a 17 digit long division problem. They would never ask you to do this. If the repeat length were 18, again you couldn't say anything other than that the 53rd digit would also be the same as the 17th. If the repeat length were 12 or 6, then the 53rd digit would be the same as the 5th. If the repeat length was 9 then the 53rd digit would be the same as the 8th. This still seems like too much division. This leaves 2, 3, 4, and 5 as possible repeat lengths. You can figure out which by just starting the long division or by figuring out whether 37 is a factor of 99, 999, 9999, or 99999. It's a factor of 999, so the repeat length is 3. This makes the 53rd digit the same as the second. The answer is 4.

The MathCounts part of this book has a section that explains repeat lengths better and tells you more about finding repeat lengths in other situations, e.g. when the denominator is not prime. I do it there instead of here because (1) this section is getting long, (2) what I've told you already should be enough to do IMLEM problems, and (3) I wanted to explain things using modular arithmetic, which we'll learn about when we get to IMLEM meet #4.

Category 5 – Algebra

The algebra topics for meet #2 are word problems with 1 unknown, working with formulas, and reasoning in number sentences.

As with many of the Algebra categories there's not much I can tell you in these notes. The problems usually aren't hard and most don't have tricks to them. You do need to be good at Algebra. Some problems also require fairly involved calculations in which it's easy to make an arithmetic error.

As always, problems are easier if you've worked through similar problems recently. The examples below cover most of what's been on past meets. I also include discussions of the few tricks that would make some of the problems easier. After you work through these notes it would be a good idea to go through some old meets for further practice.

5.1 Sums of Arithmetic Sequences

I don't know which of the stated topics this is, but one thing that seems to come up regularly is problems involving the sum of an arithmetic sequence. For example,

Question: The sum of five consecutive multiples of seven is 175. Find the product of the smallest and largest of these numbers.

The best thing to remember in problems like this is that the sum of an arithmetic sequence is equal to the number of terms times the middle term. If five terms sum to 175, then the middle term is 175/5 = 35. From here you can write down that the sequence is 21, 28, 35, 42, 49. The answer is 21 × 49 = 1029.

This problem didn't really need any algebra. If you did want to do it with algebra the trick to making the problem easier is to let the middle term in the sequence be x. This way the algebra is $(x-14) + (x-7) + x + (x+7) + (x+14) = 175$, which immediately simplifies to $5x = 175$. If you'd instead used x for the first term in the sequence the equation to find the first term would have looked like $x + (x+7) + (x+14) + (x+21) + (x+28) = 175$. It's not particularly hard to solve this problem, but adding up 7+14+21+28 and then subtracting the result from 175 would take longer and give you more opportunities to make an adding mistake.

Here's an example of similar but harder problem combining algebra and number theory.

Question: The sum of 3 consecutive positive numbers that are multiples of eleven is a multiple of ten. What is the smallest possible value for the product of the three numbers?

The answer is 1317690. This problem also illustrates my comment that they sometimes write problems that require computations that can take a while and lead to arithmetic errors.

5.2 Reasoning in Number Sentences

Reasoning in number sentences seems to mean answering questions like:

Question: The number that is eight less than three times the sum of a number and 7 is four more than twice the difference when the number is subtracted from 22. Find the number.

To answer questions like this you just need to get used to translating them to algebra. This question is asking you to find x in the equation:

$3(x+7) - 8 = 2(22-x) + 4.$

Collecting all the x's on the left side and the constants on the right this simplifies to $5x=35$, so the answer is 7.

Another example to try for practice is

Question: Five times the sum of a number and eight is ten less than the product of four and the sum of five and twice the number. What is the number?

The answer to this one is 10.

5.3 Working with Formulas

Working with formulas appears to mean plugging numbers into formulas. For example, they could ask:

Question: The formula for the circumference of a circle of radius r is $2\pi r$. What is the circumference of a circle of radius 5cm. Use 3.14 for π and round your answer to the nearest tenth of a centimeter.

This is easier than most questions they've asked but the idea is the same: you just plug in 5 where you see the r in the formula and get $10\pi \approx 31.4$cm.

To make these problems harder they typically either ask you to plug in a few things or give problems that require tedious calculations, like computing $(3.14)^2$. Here's an example of a problem with a few things to plug in:

Question: The volume of a pyramid is given by $A = \frac{1}{3} bh$, where b is the area of the base of the pyramid and h is the height of the pyramid. A tetrahedron is a pyramid whose base is an equilateral triangle and whose sides are all equal. The formula for the area of an *equilateral triangle with side-length s is $s^2\sqrt{3}/4$. The height of a tetrahedron with side-length s is $s\sqrt{6}/3$. Find the volume of a tetrahedron with side-length 2.*

This problem is just asking you to compute $\frac{1}{3}(s^2\sqrt{3}/4)(s\sqrt{6}/3)$ for $s=2$. The easiest way to do it is probably to first simplify the formula to $s^3\sqrt{2}/12$ and then plug in to get $2\sqrt{2}/3$.

Another thing they could done in the problem above is given you the first two formulas, then told you the formula for the volume of a tetrahedron, and then asked you to find the formula for the height. Here's another problem that works in a similar way. It's probably harder than anything they'll ask in IMLEM.

Question: The area of a regular pentagon with side s is $\frac{1}{4}s^2\sqrt{25+10\sqrt{5}}$. A formula for the area of any regular n-gon with side s is $A = \frac{1}{4}ns\sqrt{4R^2-s^2}$, where R is the radius of the circumscribed circle. Use these formulas to find the radius of the circle that circumscribes a regular pentagon with side-length one.

This question asks you to plug in $n=5$ and $s=1$, and then solve for R in the equation

$$\frac{1}{4}s^2\sqrt{25+10\sqrt{5}} = \frac{1}{4}ns\sqrt{4R^2-s^2}$$

The answer is $\frac{1}{10}\sqrt{50+10\sqrt{5}}$.

5.4 Word Problems with One Unknown

Some problems in this category ask you to solve problems like

Question: Kate is eight and a half years younger than Caroline. In five years Kate will be half as old as Caroline. How old is Kate?

The answer to this question is three-and-a-half. Write k for Kate's age. The first sentence tells you that Caroline's age is $k + 8\frac{1}{2}$. The second says that $k + 5 = \frac{1}{2}(k + 8\frac{1}{2} + 5)$. You solve this to get the answer. (The easiest way is to first multiply both sides by 2.)

My solution to this problem illustrates an important thing to keep in mind: the category is "word problems with one unknown", not "word problems with more than one unknown". One could have solved the problem above by writing down two equations:

$c = k + 8\frac{1}{2}$

$k+5 = \frac{1}{2}(c+5)$

In this case, this wouldn't have been much harder. What's important to remember though, is that solving multiple equations is never necessary. If a problem appears to require this and looks hard, then you should look for a different solution method that involves solving one equation at a time. For example, consider

> *Question: Each shape in the picture below represents a number. It represents the same number every time it appears. If the square represents a negative number, what is the value of the triangle?*

$$\triangle^3 = \lozenge + \lozenge$$
$$\square \times \square + \circ = 3$$
$$\lozenge \times \square + 3 = \circ$$
$$\circ - 3 = 2(\circ + 5)$$

You could try to set this up as a four-equations-in-four-unknowns problems, but these can take a while to solve by brute force even if they are linear (and this one isn't linear so there's no general way to go about solving it). The key to finding the solution is to remember that there must be a way to do it without ever solving multiple equations in multiple unknowns. This tells you that you should look for some equation you can solve right away.

The last one involves only circles. Solve that one first to find that the circle is -13. Now what can you do? The second equation only involves circles and squares, so you can solve it to find that the square is -4. Once you know the values for square and circle, the only unknown is the third equation is the diamond. Hence, you can solve that equation to find that the diamond is 4. Finally, the first equation gives that the triangle is 2.

IMLEM Meet #3

Meet #3 marks a shift in the IMLEM season. Most students will find that the last three meets cover more unfamiliar topics than did the first two. This can make the meet difficult. In one not-so-distant past year the Bigelow team only scored 42 points. That's right, 42, for ten individuals plus the team round.

I think IMLEM recognizes that a shift occurs here. The reason I say this is that there's another change in meet #3: there's less variety and the questions are more predictable from year to year. From my perspective it makes the meet a little boring. It also will make this packet less about math and more about old problems than I'd like it to be. It does, however, make the meet easier to prepare for. It can also bring in some team strategy: it may make sense for newcomers to focus on learning the material in just a couple categories.

You should think of this meet as presenting a big opportunity. Sure, your scores may be lower than in the first two meets. But if you do keep up your scores close to where they were in the earlier meets you could build up a nice lead over the teams that don't.

Category 2 – Geometry

The stated topics for meet #3 are properties of polygons and the Pythagorean Theorem. In past years, a number of the questions have been on topics also covered in meets #1 and #2. I'll give brief reviews of the older topics and then discuss the new ones.

2.1 Review of Meet #1: Angles in Polygons

The main things to remember about polygons from meet #1 are:

- The angles in any n-gon add up to $(n-2)\,180°$. For example, the angles in any 12-gon add up to 1800°.

- A corollary is that the measure of each angle in a regular n-gon is $((n-2)\,/\,n)\,180°$. For example, each angle in a regular octagon is $(6/8) \times 180 = 135°$.

- The n exterior angles of an n-gon add up to 360°. An *exterior angle* is the angle that a polygon makes when you put it on top of a table. The diagram below shows two examples:

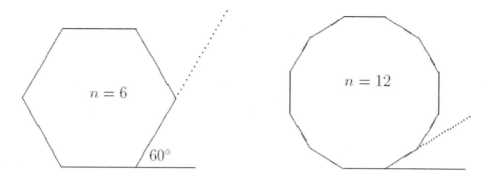

In the one on the left the exterior angle is 60°. In the one on the right it's 30°.

- In a regular *n*-gon, the measure of each exterior angle must be 360/*n*. For example, in a hexagon (6-gon) each exterior angle is 60°. Interior and exterior angles are supplements (remember what that means?), so the interior angle is 120°. In a dodecagon (12-gon), each of the exterior angles is 30°. The interior angles are 150°.

Meet #3 has included some questions just like those on meet #1 in which the answer is very easy if you know these formulas. For example,

Question: (IMLEM Meet #3, Jan. 2006) How many degrees are in the interior angle of a regular 15-gon.

Answer: 156°. (13/15) 180 = 13×12 = 156. Another way to do this is to say that the exterior angle is 360/15 = 24, so the interior angle must be 180 – 24 = 156.

Another example is

Question: (IMLEM Meet #3, Jan. 2002) The exterior angle of a regular n-gon measures 15 degrees. How many sides does the polygon have?

Answer: 24. The exterior angle formula tells us that 360/n=15. A (very) little algebra gives n=360/15=24.

The questions above used only a very slight bit of algebra. The meet has also included a few questions that use a little bit more algebra. For example,

Question: (IMLEM Meet #3, Jan. 2002) The interior angle of a regular polygon is five times as great as an exterior angle of the same polygon. How many sides does the polygon have?

Answer: 12. One way to do this is by solving the equation 5× (360/n) = ((n-2) / n) 180. This isn't as hard as it looks at first because the n's cancel and you can divide both sides by 180.

An alternate way to do the above problem is with a two step approach. First, find the exterior angle. You can do this by solving $5x = 180 - x$. The answer is 30. Then solve the problem: "The exterior angle of a regular n-gon is 30 degrees. Find n." Yet another way to solve the problem would be to memorize the list of n-gon angle measures I gave in the meet #1 packet. You could then just go down the list and see which had the interior angle equal to five times the exterior angle.

Here's the list again.

n	Popular name for n-gon	Sum of interior angles	Interior angle in regular n-gon	Exterior angle in regular n-gon
--	-------	-------	-------	-------
3	Triangle	180	60	120
4	Quadrilateral	360	90	90
5	Pentagon	540	108	72
6	Hexagon	720	120	60
8	Octagon	1080	135	45
10	Decagon	1440	144	36
12	Dodecagon	1800	150	30
15	15-gon	2340	156	24
18	18-gon	2880	160	20
20	20-gon	3240	162	18
n	n-gon	$180(n-2)$	$(n-2)/n \cdot 180$	$360/n$

2.2 Review of Meet #2: Areas of Polygons

The meet #2 packet gave lots of area formulas. A few of the easier ones seem to come up again on meet #3. They are:

- The area of a triangle with base b and height h is $A = \frac{1}{2} b h$. Note that in a right triangle if you take either of the legs to be the base, then the other is the height.

- The area of a square with side s is $A = s^2$.

- The area of a rectangle with length ℓ and width w is $A = \ell w$.

- The area of a trapezoid with bases a and b and height h is $A = \frac{1}{2}(a + b) h$.

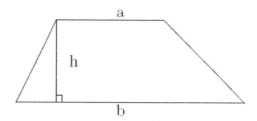

- The area of a parallelogram with base b and height h is $A = b\,h$.

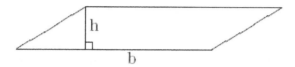

Usually the way in which these come up on meet #3 is that you have to go back and forth between facts about areas and facts about perimeters to find an answer. A very easy but otherwise not atypical example would be a question in which they told you that the area of a square was 100 and asked you to find the perimeter.

The problem below is a slightly more involved example in which you need to use two area formulas:

Question: (IMELM Meet #3, Jan. 2001) How many inches are in the side length of a square that has half the area of the parallelogram shown below?

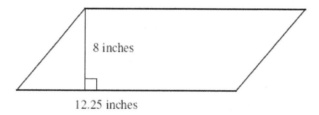

First, you use the parallelogram formula to find that the area of the parallelogram is $8 \times 12.25 = 98$. Then you use the square formula to say that the side-length of a square with area 49 is 7.

The problem below is a more difficult and clever example.

Question: (IMELM Meet #3, Jan. 2001) Triangle PYTH is a right triangle with PY=6cm and PH=8cm. How many centimeters long is PT, which is an altitude relative to side YH?

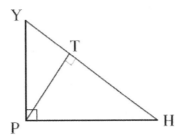

The key to the problem is to recognize that you can find the area in two different ways. One is to use ½ base × height formula with the two legs as the base and the height, i.e. A = ½ PY PH = 24. The other is to use the ½ base × height formula with the hypotenuse YH as the base and PT as the height. The Pythagorean theorem gives that the length of the hypotenuse is 10. Hence, the area is also ½ 10 PT. The answer is 24/5.

Problems based on this one seem to come up all the time on various math contests, so it's a good one to remember.

2.3 Diagonals in Polygons

One thing that seems to come up a lot in these meets is the number of diagonals that a polygon has. A *diagonal* is a line segment that connects two nonadjacent vertices of a polygon. You're probably used to seeing these in a rectangle, where they're the two diagonal lines connecting opposite corners. In other polygons there are more diagonals. For example, in the pentagon below each of the dotted lines is a diagonal

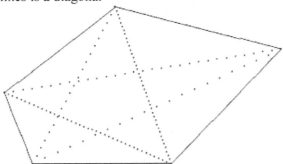

The main fact about diagonals that has come up over and over again on IMLEM meets is

- An *n*-gon has ½ *n* (*n* – 3) diagonals.

For example, the formula says that a pentagon has ½ × 5 × 2 = 5 diagonals. The reason why the formula works is that you can draw all the diagonals by picking each of the *n* vertices and connecting it to *n* – 3 others. The only three that don't work are the vertex you picked and its two neighbors. This might make you think the answer should be $n(n-3)$ but it's not. If you follow this procedure you'd draw each diagonal twice: once starting from one end and once starting from the opposite end.

Some past meet #3 questions are very easy if you know this formula. For example,

Question: (IMLEM Meet #3, Jan. 2004) How many diagonals can be drawn in a decagon. (Note: A decagon is a polygon with 10 sides.)

Question: (IMLEM Meet #3, Jan. 2002) How many diagonals can be drawn in the dodecagon shown at right.

Question: (IMLEM Meet #3, Jan. 2003) How many more diagonals can be drawn in a nonagon (a polygon with nine sides) than in a septagon (a polygon with 7 sides).

The answers are 35, 54, and 13. One thing to note is that they sometimes tell you what words like "decagon" mean and sometimes don't. (In the dodecagon problem they did include a picture of the dodecagon so you could have counted the number of sides.) For this reason you may want to try to learn the names for the n-gons I gave earlier.

Again, they also seem to be willing to put on problems that require some use of algebra. For example, they might ask:

Question: A certain polygon has twice as many diagonals as sides. How many sides does it have?

The problem tells you that $\frac{1}{2} n (n-3) = 2n$. Cancelling the n's from both sides gives $\frac{1}{2} (n-3) = 2$. The answer is 7. You also could have found this by making a table.

2.4 Pythagorean Theorem

The Pythagorean Theorem says that the sides of a right triangle satisfy $a^2 + b^2 = c^2$. The diagram below illustrates a nice proof of the theorem. The length of each side of the big square is $a + b$, so the area of the big square is $(a + b)^2 = a^2 + 2ab + b^2$. Another way to find the area of the big square is as the sum of the four right triangles and the central square. The four right triangles each have area $\frac{1}{2} ab$, so we have $a^2 + 2ab + b^2 = 4 (\frac{1}{2} ab) + c^2$. Cancelling the $2ab$ terms from both sides gives the Pythagorean Theorem.

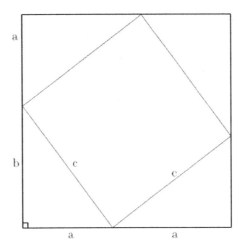

2.4.1 Some Pythagorean triples

When doing problems involving the Pythagorean theorem it helps to be able to recognize common Pythagorean triples. For example, if you see a long skinny right triangle in a figure and see that the short side has length seven it's a good guess that the triangle is a 7-24-25 triangle because there aren't any other right triangles that have a seven in them.

The most common right triangle is a 3-4-5 triangle. Also popular on math meets are triangles that are scaled up versions of this one: 6-8-10, 9-12-15, 12-16-25, 15-20-25, etc.

Every odd number is the smallest number in a Pythagorean triples. A few of these (with a few multiples) are:

5-12-13   (10-24-26, 15-36-39, etc.)
7-24-25   (14-48-50, etc.)
9-40-41
11-60-61

All of these work exactly the same way. If $a^2 + b^2 = c^2$, then $a^2 = c^2 - b^2 = (c - b)(c + b)$. You can make a triple for any odd number by having $c - b = 1$ and $c + b = a^2$. (Challenge 1: Why does this only work for a odd? Challenge 2: Why is there only one right triangle with one leg of length a and the others of integer length if a is prime?) For other a's there can be multiple Pythagorean triples.

A couple other Pythagorean triples that aren't multiples of those above are:

8-15-17
20-21-29

2.4.2 IMLEM Pythagorean problems

Most IMLEM Pythagorean problems seem to ask you to find the perimeter of a complicated polygon. Where the Pythagorean theorem usually comes in is that some of the edges in the polygon are hypotenuses of right triangles. You first do something to find the lengths of the sides of the triangles and then use the Pythagorean theorem to find the lengths of sides you need to compute the perimeter. One example is:

Question: (IMLEM Meet #3, Jan. 2006) Three right triangles are joined together to form the concave pentagon shown below. The measure of GA is 15 units, AS is 20 units, SM is 21 units and ME is 35 units. How many units are in the perimeter of pentagon GAMES?

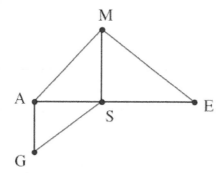

Here's one way to do this one. First use the fact that GAS is a right triangle to find GS=25 (Note that 15-20-25 is a 3-4-5 multiple). Next, use the fact that MSE is a right triangle to find SE=28. (Again 21-28-35 is a 3-4-5 multiple.) Finally, we have $MA^2 = 20^2 + 21^2 = 841 = 29^2$. The perimeter of GAMES is 15+29+35+28+25 = 132.

The problem above was easier than many IMLEM problems because you didn't need to do anything other than use the Pythagorean theorem a few times. In some other problems you also need to use your area facts and sometimes need to fill in lengths other than those in the perimeter to get the answer. A typical example is:

3. In the figure at right, segment YG measures 8 units, angle PGY is a right angle, square THAY has an area of 100 square units, angle OGA is a right angle, PGRO is a square, and triangle PGY has an area of 60 square units. How many units are in the perimeter of polygon PYTHAGOR?

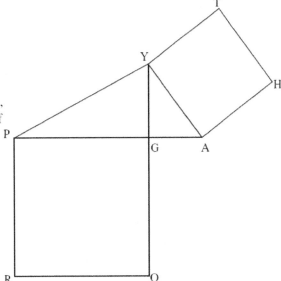

Answers

1. _____

2.

The answer to this one is 98.

2.4.3 The distance formula

Consider a set of points with locations in standard (x, y) coordinates. The distance from the point $(0,0)$ to the point (x,y) is $\sqrt{x^2 + y^2}$. This is an easy consequence of the Pythagorean theorem. Another easy consequence is that the distance between (x_1,y_1) and (x_2,y_2) is $\sqrt{(x_1 - x_2)^2 + (y_1 - y_2)^2}$.

A similar formula works in three (or more) dimensions. The distance between (x_1,y_1,z_1) and (x_2,y_2,z_2) is $\sqrt{(x_1 - x_2)^2 + (y_1 - y_2)^2 + (z_1 - z_2)^2}$. Proving that this works is a nice exercise. The way to do it is to draw a line in the x-y plane connecting (x_1,y_1,z_1) and (x_2,y_2,z_1), find its length, and then look at the right triangle with this and the vertical segment connecting (x_2,y_2,z_1) and (x_2,y_2,z_2) as its two legs.

2.4.4 Problem solving tip: don't calculate more square roots than you need to

One thing you can do to solve some IMLEM problems more quickly is to put off taking square roots until it is really necessary. In fact, some problems seem to be designed to take a very long time if you don't know this trick and to be easy if you do. Here's one example:

Question (IMLEM Meet #3, Jan. 2005) In the figure below triangles ABC, ACD, and ADE are right triangles and AB, BC, CD, and DE have the same measure. If the measure of AB is 2cm, how many centimeters are in the measure of side AE.

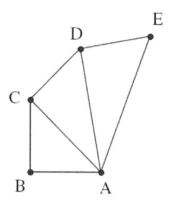

To solve this problem you need to first find AC, then AD, and then AE. It's not so hard that you can't calculate them all explicitly. First find AC = $\sqrt{2^2 + 2^2} = \sqrt{8} = 2\sqrt{2}$. Then, compute AD = $\sqrt{2^2 + (2\sqrt{2})^2} = \sqrt{4+8} = \sqrt{12} = 3\sqrt{2}$, etc. This really is more work than is necessary, however. When computing AC it's better to stop at AC = $\sqrt{2^2 + 2^2}$. Then, at the next step write AD = $\sqrt{2^2 + (\sqrt{2^2 + 2^2})^2} = \sqrt{2^2 + 2^2 + 2^2}$ and stop there. Only at the end when you find AE = $\sqrt{2^2 + 2^2 + 2^2 + 2^2}$ should you bother to compute the square root.

Another example of a question with a similar trick is

Question: (IMLEM Meet #3, Jan. 2001): Mr. Snood packed his homing pigeon, Homer, in the car one Saturday morning and set out for a drive. He drove 7 miles due west, then 6 miles due south, then 14 miles due west, then 4 miles due south, and finally 3 more miles west. He then released Homer from his cage and the bird flew off toward home. How many miles will Homer have to fly if he flies directly back to the house.

Here, you might be tempted to start drawing lots of lines connecting the origin to different intermediate points on his route. Don't do it. The way to do this problem is to start by saying the ends up 24 miles west and 10 miles south of where he started. The distance to where he started is $\sqrt{10^2 + 24^2} = 26$.

Category 3 – Number Theory

The number theory category in meet #3 covers scientific notation and bases. Preparing for this topic could take a fair amount of studying, but it's a good set of things to know and I personally think the bases topic is one of the more interesting things they do all year. If you do study (and are good at arithmetic) you should do fine.

3.1 Scientific Notation

In *scientific notation* numbers are written as products of the form $a \times 10^n$, where a is a number satisfying $1 \leq a < 10$ and n is an integer (which can be positive or negative or zero).

A few examples are:

$12,340 = 1.234 \times 10^4$
$7 = 7 \times 10^0$
$0.5 = 5 \times 10^{-1}$
$0.0037 = 3.7 \times 10^{-3}$

There is a unique way to write every number (other than zero) in scientific notation.

3.1.1 Improper scientific notation

The most common mistake people make when doing scientific notation is to forget the requirement that $1 \leq a < 10$. To drill this in, let me note that the following numbers are NOT in scientific notation:

13.4×10^{-7}
0.4×10^6

The proper way to write these numbers in scientific notation is 1.34×10^{-6} and 4×10^5. A good way to keep the mechanics of converting from improper scientific notation to proper scientific notation straight is to think of adding one to the exponent every time you move the decimal point to the left and subtracting one from the exponent every time you move the decimal point to the right.

3.1.2 Multiplying in scientific notation

Multiplying in scientific notation is a three step process: (1) regroup the terms; (2) do the multiplications; and (3) convert the improper scientific notation back to proper scientific notation (if necessary).

For example to multiply $(3.2 \times 10^6) \times (4 \times 10^{-3})$, regroup the terms as

$(3.2 \times 4) \times (10^6 \times 10^{-3})$,

then multiply to get (remember that you add exponents when multiplying powers of ten)

12.8×10^3,

and then standardize to get 1.28×10^4.

3.1.3 Dividing in scientific notation

Dividing in scientific notation is similar. For example, to divide $(3.2 \times 10^6) \div (4 \times 10^{-3})$, I first regroup the terms as

$(3.2 / 4) \times (10^6 / 10^{-3})$,

Then do each of the divisions to get

0.8×10^9

And then standardize this to get 8×10^8.

3.2 IMLEM Scientific Notation Problems

A convenient aspect of this category is that they ask essentially the same scientific notation question every year. Also convenient is that the same two-step trick makes such questions much simpler.

Question: Simplify the expression below. Write your result in scientific notation.

$$\frac{(5.4 \times 10^{13}) \cdot (1.4 \times 10^{-6})}{(3.5 \times 10^{-14}) \cdot (6 \times 10^{19})}$$

The two-step trick to these problems is (1) first write them in improper scientific notation so there are no decimal points, and then (2) cancel common factors before doing any of the multiplications or divisions.

For example, in the problem above you first write it as

$$\frac{(54 \times 10^{12}) \cdot (14 \times 10^{-7})}{(35 \times 10^{-15}) \cdot (6 \times 10^{19})}$$

Regrouping, this is

$$\frac{54 \times 14 \times 10^5}{35 \times 6 \times 10^4}$$

We then note that $54 = 6 \times 9$, $14 = 2 \times 7$, and $35 = 5 \times 7$. Hence, we can cancel both a six and a seven from the numerator and denominator. This leaves,

$$\frac{18 \times 10^5}{5 \times 10^4}$$

which simplifies to 3.6×10^1.

Here are a couple more you can do for practice if you like:

Question: Simplify the expression below. Write your result in scientific notation.

$$\frac{(1.11 \times 10^4) \cdot (6 \times 10^{-2})}{(2.5 \times 10^8) \cdot (7.4 \times 10^6)} \quad \text{and} \quad \frac{(1.47 \times 10^{-5}) \cdot (1.44 \times 10^{23})}{(1.96 \times 10^9) \cdot (3 \times 10^{-5})}$$

The answers are 3.6×10^{-13} and 3.6×10^{13}.

3.3 Basics of Bases

Our standard number system is sometimes called the *base 10* number system. This means that the place values in a number correspond to powers of ten. For example, when we write 127 what we mean is $(1 \times 10^2) + (2 \times 10^1) + 7$.

Why do we use base 10 numbers? Part of the answer is that a system with place values makes adding, subtracting, multiplying and dividing much easier than it was in Roman numerals. Another part, however, is that we have ten fingers.

Those of you who watch *The Simpsons* will probably have noticed that characters on the show only have eight fingers. How would we do math if we only had eight fingers? Most people think the answer is that we would use a *base 8* number system. In the base 8 number system there are only eight digits: 0, 1, 2, 3, 4, 5, 6, and 7. Place values represent powers of eight. For example, the number 127 would mean $(1 \times 8^2) + (2 \times 8^1) + 7$, i.e. the number we call 87.

Things can get confusing when you start having numbers in different bases. The standard way to keep this straight is to write the base as a subscript immediately after a number. For example, we can write $127_{\text{base }8}$ or 127_8 for the base 8 number and $127_{\text{base }10}$ or 127_{10} for the number we're used to calling one hundred and twenty seven.

3.3.1 Converting to base 10

One standard thing you need to know how to do is convert numbers written in different bases to base 10. You do this by just writing down what the number means and then doing all of the exponentiations, multiplications and additions. For example,

Question: What is the base 10 value of the base 4 number 3021?

The way you do this is to write $3021_{base\ 4} = (3 \times 4^3) + (2 \times 4^1) + 1 = (3 \times 64) + (2 \times 4) + 1 = 201_{base\ 10}$. If they ask the question this way you can just write 201 for the answer.

3.3.2 Converting from base 10 to base b

Another standard thing you need to do is to convert numbers from base 10 to other bases. For example, you could get asked:

Question: Express the base 10 number 691 as a base 8 number.

The first thing you should do to solve such problems is to write down the first few powers of 8: 1, 8, 64, and 512. Then, ask yourself how many 512's fit into the number 691. The answer is 1 with a remainder of 179. Next, ask yourself how many 64's fit into the number 179. The answer is 2 with a remainder of 51. Now, 51 divided by 8 is 6 with a remainder of 3. Hence, we've found that $691 = (1 \times 512) + (2 \times 64) + (6 \times 8) + 3 = (1 \times 8^3) + (2 \times 8^2) + (6 \times 8^1) + 3 = 1263_8$.

Here's another practice question:

Question: Express the base 10 number 500 as a base 3 number.

The answer is $200112_{base\ 3}$.

3.3.3 Bases larger than 10

When working in base 4 we only ever use the digits 0, 1, 2, and 3. This is not a problem (provided you're not worried about the other digits feeling left out.) When working with bases greater than ten a new problem arises: we need a single symbol that we can use to mean ten, eleven, etc. The standard way to do this is to use the letter A as the ten symbol, B as the eleven symbol, etc. For example, suppose you want to write $171_{base\ 10}$ as a base 16 number. Following the method above you find $171 = (10 \times 16) + 11$. Hence, the standard way to write 171 in base 16 is $AB_{base\ 16}$.

Here's a practice question:

Question: In base 16 it is standard to use 0, 1, 2, 3, 4, 5, 6, 7, 8, 9, A, B, C, D, E, and F as the digits. Express the base 16 number 2FE as a base 10 number.

The answer is $(2 \times 256) + (15 \times 16) + 14 = 766_{base\ 10}$.

3.4 Converting From Base A to Base B

A number of IMLEM questions ask you to convert between two different bases. For example,

Question: Express the base 3 number 2111 as a base 9 number.

3.4.1 The standard method: go through base 10

You already know one way to do problems like this: you could convert the base 3 number 2111 to a base 10 number (the answer is $67_{base\ 10}$), and then convert $67_{base\ 10}$ to a base 9 number. This method works fine for small numbers like this, but can take a while and gives you multiple chances to make arithmetic errors.

When the numbers get big the above method is hopeless. Consider, for example,

Question: Express the base 3 number 20211002112021 as a base 9 number.

To use the above method you'd need to start by figuring out all the powers of 3 up to 3^{13} (which is 1594323) then add up a bunch of 7 digit numbers.

3.4.2 Direct conversions between base b and base b^2

A better way to do the two problems above is to convert directly between the bases without ever using base 10. Here's how this works:

The base 3 number 2111 means $(2 \times 3^3) + (1 \times 3^2) + (1 \times 3^1) + 1$. Group these terms into pairs. The first two are: $(2 \times 3^3) + (1 \times 3^2) = (2 \times 3 \times 3^2) + (1 \times 3^2) = (2 \times 3 + 1) \times 3^2$. The next two are: $(1 \times 3) + 1$. Hence, we've shown that $2111_{base\ 3} = (2 \times 3 + 1) \times 3^2 + (1 \times 3 + 1) = 74_{base\ 9}$

This may seem complicated the first time you see it, but once you get used to the pattern it's quite easy. You just group the digits into pairs and convert each pair from base 3 to base 9. For example, the way I'd do the harder problem above is to write.

$$20211002112021_{base\ 3} = 20\ 21\ 10\ 02\ 11\ 20\ 21_{base\ 3}$$
$$= 6\ \ 7\ \ 3\ \ 2\ \ 4\ \ 6\ \ 7_{base9}$$
$$= 6732467_{base9}$$

The first line is just putting spaces between the pairs (be sure to start from the right if the number has an odd number of digits). The second is a bunch of easy base 3 to base 9 conversions, e.g. $20_{base3} = 2 \times 3 + 0 = 6_{base\ 10} = 6_{base9}$. and $21_{base3} = 2 \times 3 + 1 = 7_{base\ 10} = 7_{base9}$. The third line is just deleting the extra spaces.

Converting from base 9 to base 3 is just as easy. For example,

$$5134_{base9} = 12\ 01\ 10\ 11_{base3}$$
$$= 12011011_{base3}.$$

The first line is just a bunch of easy base 9 to base 3 problems, e.g $5_{base9} = 5_{base10} = 12_{base\ 3}$. The second is taking out the spaces.

Here's one example from a real IMLEM meet:

Question: (IMLEM Meet #3, January 2006) Express the base 3 number 121212 as a base 9 number.

The answer is $555_{base\ 9}$.

3.4.3 Direct conversions between base q^m and base q^n

In the previous section one base was the square of the other. The "square" part is why I grouped the digits in groups of 2. If one base was the cube of the other, things would be just as easy. All you'd need to do is to use groups of three. For example,

Question: Express the binary number 10100101 as a base 8 number.

"Binary" means base 2. Hence, we convert this number as:

$$10100101_{base\ 2} = 10\ 100\ 101_{base\ 2}$$
$$= 2\quad 4\quad 5_{base8}$$
$$= 245_{base8}$$

In other problems, one base isn't any integer power of another, but both are powers of the same number. For example, consider

Question: (IMLEM Meet #3, January 2001) Express the base 4 number 3132 as a base 8 number.

There are two different variants of the above method that can be used in such a situation. Both are quicker than going through base 10 if you're used to them, but they can also be confusing so you'll have to decide for yourself whether you want to try them.

One method for doing this problem without going through base 10 is to go through base 2. Here's how this one would look:

$$3132_{base\ 4} = 11\ 01\ 11\ 10_{base\ 2}$$
$$= 11\quad 011\quad 110_{base\ 2}$$
$$= 3\quad 3\quad 6_{base8}$$
$$= 336_{base8}$$

The other method for doing this problem is to group the digits in blocks of length 3 (which is the smallest number such that 4^n is a power of 8.) You then do the conversion block by block using any method you like (e.g. the going through base 10 method).

The way this would look (in a slightly harder problem) is
$$1003132_{\text{base 4}} = 1\ 003\ 132_{\text{base 4}}$$
$$= 1\ 03\ 36_{\text{base 8}}$$
$$= 10336_{\text{base8}}$$

The first line is adding spaces every three digits. The second is converting base 4 numbers to base 8 numbers. Here, $1_{\text{base 4}} = 1_{\text{base 8}}$ and $003_{\text{base4}} = 03_{\text{base8}}$ are obvious. I found $132_{\text{base4}} = 36_{\text{base8}}$ by going through base 10. One tricky thing to keep in mind is that you need to write all base 8 numbers here as two digit numbers, even if it's 03.

You'll have to decide for yourself whether you think either of these methods is easier and less mistake-prone than just going through base 10.

One other thing to keep in mind is that methods in this subsection only work if the two bases are both powers of the same number. If you're asked to convert from base 4 to base 7 or from base 8 to base 9 you must use the standard method of just going through base 10.

3.5 Arithmetic in Other Bases

One final type of question that has come up is questions asking you to do simple arithmetic problems in other bases. Consider for example:

Question: Find the sum of the two base 5 numbers 2122 and 1401. Express your answer as a base 5 number.

One way to do questions like this is straightforward: convert each number to base 10, add the two numbers together, and then convert the result to base 5. Again, this is a pretty safe method. The one drawback is that it can take a while if the numbers are big.

3.5.1 Addition in base b

The alternate way to do questions like this is just to do the addition in base 5 without ever converting to base 10. Pretty much everything you're used to doing in base 10 works in other bases too. You just have to keep in mind that the base is 5 instead of 10. For example,

```
  2 1 2 2
+ 1 4 0 1
----------
  4 0 2 3
```

The way you do this is to just do the addition from right to left as you were taught in first grade. In the rightmost two columns we have $2 + 1 = 3$ and $2 + 0 = 2$. In the third column we have $4 + 1$. There is no digit 5 in the base 5 number system, so the answer is $4 + 1 = 10_{\text{base } 5}$. What you do when this happens is to write down zero and carry a 1 to the next column. In the left column you then add $1 + 2 + 1$ to get 4.

A couple of other problems you could do for practice are

Question: Find the sum of the two base 8 numbers 2162 and 1441. Express your answer as a base 8 number.

Question: Find the sum of the two base 6 numbers 215 and 341. Express your answer as a base 6 number.

The answers are 3623_{base8} and 1000_{base6}.

3.5.2 Subtraction in base b

Subtraction works similarly. It's really simple in a question like:

Question: Find the result when the base 8 number 124 is subtracted from the base 8 number 2375. Express your answer as a base 8 number.

In a question like the one below you need to use borrowing:

Question: (IMLEM Meet #3, January 2004) Solve the following base 5 equation for n:

$$2343_{\text{base5}} + n_{\text{base5}} = 4032_{\text{base 5}}$$

Express your answer as a base 5 number.

This again, however, works just like it did in second grade. You just need to remember that whenever you borrow you only get 5 instead of 10. For example the answer to the problem above is

```
  4 0 3 2
- 2 3 4 3
----------
  1 1 3 4
```

To do this, you borrow over and over again. First $2 - 3$ would be negative. So you borrow 5 and change the 3 in the top second column to a 2. $(5 + 2 - 3) = 4$. So I start by writing a 4 in the rightmost column of the answer. Next, $2 - 4$ would be negative, so you borrow 5 and change the 0 in the top number to a -1. $(5 + 2) - 4 = 3$, so you write a three in the next column of the answer. Again borrowing 5 we get $(5 + -1) - 3 = 1$, so we just write down a 1. Finally, there's no borrowing needed $3 - 2 = 1$.

For practice a slightly easier problem is:

Question: Find the result when the base 8 number 1024 is subtracted from the base 8 number 5333. Express your answer as a base 8 number.

The answer is $4307_{base\ 8}$.

3.5.3 Multiplication and division in base b

You can also do multiplication and division in other bases just like you do in base 10. It can be confusing, however, so I wouldn't normally recommend it unless the problem has something special about it that makes it easy to do the multiplication in base b.

One example where you obviously want to do it this way is:

Question: Find the base eight product of the two base 8 numbers 3741 and 10. In other words, evaluate $3741_{base\ 8} \times 10_{base8}$ in base 8.

The answer is $37410_{base\ 8}$. Just like you can multiply by ten in base 10 by adding a zero to the end of a number you can multiply by eight in base 8 by adding a zero to the end of a number.

If the problem is not too much more complicated you can also do it by this method. For example, multiplying out $3741_{base\ 8} \times 11_{base8}$ looks like

```
    3 7 4 1
 ×     1 1
-----------
    3 7 4 1
  3 7 4 1
-----------
  4 3 3 5 1
```

The first two lines of this are just putting the numbers down in the correct place values like you always do the in multiplication. You then do the addition as a base 8 addition problem.

Long division works too. In easy problems it's not at all bad. For example, $4462_{base\ 7} \div 2_{base\ 7}$ is $2231_{base\ 7}$. You could also do $43351_{base\ 8} \div 11_{base\ 8}$ as

```
       3 7 4 1
 1 1 ) 4 3 3 5 1
       3 3
       1 0 3
         7 7
           4 5
           4 4
```

$$1\ 1$$
$$\underline{1\ 1}$$
$$0$$

I wouldn't recommend doing this on a meet though unless you've had quite a bit of practice.

3.6 Word Problems Related to Bases

The meet also has had a few word problems related to bases. There isn't much special that you need to know to do them, but I thought it might be useful to have seen the types of problems they ask, so here are two:

1. A set of base four blocks includes unit cubes, longs, flats, and blocks. Each unit cube is 1 cm by 1 cm by 1 cm. Each "long" is 1 cm by 1 cm by 4 cm. Each "flat" is 1 cm by 4 cm by 4 cm. And each "block" is 4 cm by 4 cm by 4 cm. The base four number 2023 is given by the picture below. What is the base ten value of $2023_{\text{base four}}$?

2. A box of twelve eggs is commonly refered to as a "dozen" eggs. Less common are the words "gross", which means a dozen dozen, and "great gross", which means a dozen dozen dozen. How many eggs are there in two great gross five gross nine dozen three?

3.7 Advanced Topic: Adding and Subtracting in Scientific Notation

As I said before, almost all the IMLEM scientific notation problems are multiplication/division problems where it simplifies things to cancel common factors before doing the calculations. I don't see any reason why they couldn't ask you to add or subtract instead, so you might as well learn this too.

The one thing you need to know to do addition and subtraction problems is that the first step is to convert the numbers so that they have the same exponent. For example,

$$(3.2 \times 10^6) + (2.4 \times 10^5) = (32 \times 10^5) + (2.4 \times 10^5) = (32 + 2.4) \times 10^5 = 34.4 \times 10^5 = 3.44 \times 10^6$$

You can sometimes avoid converting back at the end by putting everything in terms of the larger exponent, e.g.

$(2.7 \times 10^8) + (3.1 \times 10^6) = (2.7 \times 10^8) + (0.031 \times 10^8) = 2.731 \times 10^8$

Subtracting works similarly, e.g.

$(4.2 \times 10^{-4}) - (3.1 \times 10^{-5}) = (4.2 \times 10^{-4}) - (0.31 \times 10^{-4}) = 3.89 \times 10^{-4}$

Category 4 – Arithmetic

The arithmetic category in meet #3 covers integer powers and roots up to the sixth. It must be that they were worried that some middle school kids could understand what a fifth and sixth root was, but would be totally lost if they had to try to learn what a seventh root was. In contrast to the number theory category, this is perhaps the most boring topic of the year for me to write a packet about. The material is all very standard textbook material and there's nothing clever about the problems. I know that at Bigelow the material is scattered throughout all three years of the math curriculum, however, so I figured I should put it down in one place.

4.1 Basics of Exponents

4.1.1 Positive integer exponents

If n is a positive integer then a^n is what you get by multiplying n a's together.

For example, $3^2 = 3 \times 3 = 9$ and $2^5 = 2 \times 2 \times 2 \times 2 \times 2 = 32$.

By convention, any nonzero number raised to the zero power is one, e.g. $5^0 = 1$.

4.1.2 Negative exponents

By definition a^{-n} is the reciprocal of a^n. For example, $4^{-3} = 1/4^3 = 1/64$.

4.1.3 Raising a fraction to an exponent

A convenient thing about exponents is that to raise a fraction to an exponent you can just separately raise the numerator and denominator to an exponent:
$$\left(\frac{a}{b}\right)^n = \frac{a^n}{b^n}.$$
For example, $(5/2)^3 = 5^3/2^3 = 125/8 = 15\frac{5}{8}$. Because this is so easy, if you're asked to raise a mixed number to a power, it's almost always best to put in a/b form first. For example $(2\frac{1}{2})^3 = (5/2)^3 = 5^3/2^3 = 125/8 = 15\frac{5}{8}$. (One exception would be fractions like $(10\frac{1}{10})^2$ where you are better off using the $(a + b)^2$ formula, $(a + b)^2 = a^2 + 2ab + b^2$.)

Because $1/(a/b) = b/a$, when you raise a fraction to a negative power you can just raise the numerator and denominator to the positive power and then flip it over:
$$\left(\frac{a}{b}\right)^{-n} = \frac{b^n}{a^n}.$$
For example, $(2/7)^{-3} = 7^3/2^3 = 343/8 = 42\frac{7}{8}$.

4.1.4 Raising negative numbers to integer exponents

You can raise negative numbers to integer exponents just like you raise positive numbers to integer exponents. For example $(-2)^2 = 4$, $(-4)^3 = (-4) \times (-4) \times (-4) = -64$, $(-3/2)^3 = -27/8$, and $(-1/2)^{-4} = 16$. Note that raising a negative number to an even integer (positive or negative) will always give a positive result and raising a negative number to an odd integer will give a negative result.

4.2 Operations with Exponents

Products can be neatly simplified when the same number is raised to a power in both terms:

- $a^n \times a^m = a^{n+m}$

For example $3^2\, 3^3 = 3^5$. Sometimes you can use this rule when it's not immediately obvious by converting the numbers so that they both involve the same number raised to an exponent. For example, $9 \cdot 3^3 = 3^2\, 3^3 = 3^5$. There's nothing you can do to simplify a product like $3^2\, 5^3$.

Division works similarly (because dividing by a^m is like multiplying by a^{-m}):

- $a^n / a^m = a^{n-m}$

For example $3^5 / 3^2 = 3^3$ and $5^2/5^{-6} = 5^8$.

Raising an exponentiated number to a power is always easy: you multiply the two exponents.

- $(a^n)^m = a^{nm}$

For example $(3^5)^2 = 3^{10}$ and $(5^2)^{-1} = 5^{-2} = 1/25$.

Adding exponentiated numbers is not so easy. If the two numbers are different powers of the same number you can collect terms and use:

- $a^n + a^m = a^{n-m}\, a^m + a^m = (a^{n-m} + 1)\, a^m$

For example $3^2 + 3^4 = 3^2(1 + 3^2) = 10 \cdot 3^2 = 90$. Otherwise, there's not much to do.

To add exponentiated fractions there usually isn't anything special to know or do. You just compute each fraction, put them over a common denominator, and add. For example,

$$\left(\frac{3}{2}\right)^2 + \left(\frac{6}{5}\right)^{-1} = \frac{9}{4} + \frac{5}{6} = \frac{27}{12} + \frac{10}{12} = \frac{37}{12}.$$

4.3 Roots

You're probably all familiar with square root. The square root of 64 is 8 because $8 \times 8 = 64$. Third roots (also called cube roots) are defined similarly. The third root of 64 is 4 because $4 \times 4 \times 4 = 64$. You can probably guess how to define fourth, fifth, and sixth roots: the nth root of x is the number r such that a product of n r's is x, i.e. $r^n = x$. For example, the sixth root of 64 is 2 because $2 \times 2 \times 2 \times 2 \times 2 \times 2 = 64$.

4.3.1 n-th roots

You're probably used to seeing square roots written as $\sqrt{64} = 8$. Once we start doing n-th roots people like to avoid confusion by putting a little 2 in the square root symbol to show it's a square root, e.g. $\sqrt[2]{64} = 8$. N-th roots are then written by putting a small n in the same place. For example, the first few cube roots with integer answers are:

$$\sqrt[3]{1} = 1, \sqrt[3]{8} = 2, \sqrt[3]{27} = 3, \sqrt[3]{64} = 4, \sqrt[3]{125} = 5, \sqrt[3]{216} = 6, \sqrt[3]{343} = 7, \sqrt[3]{512} = 8, \sqrt[3]{729} = 9, \sqrt[3]{1000} = 10$$

A couple other common roots one sees are: $\sqrt[4]{16} = 2, \sqrt[4]{81} = 3, \sqrt[5]{32} = 2$, and $\sqrt[6]{64} = 2$.

One important fact about n-th roots is:

- $\sqrt[n]{a \cdot b} = \sqrt[n]{a}\sqrt[n]{b}$

This lets you find some square roots that aren't obvious, e.g. $\sqrt[2]{4900} = \sqrt[2]{49}\sqrt[2]{100} = 7 \cdot 10 = 70$. It also gets used all the time in letting you simplify square roots that don't have integer answers. For example, $\sqrt[2]{8} = \sqrt[2]{4} \cdot \sqrt[2]{2} = 2 \cdot \sqrt[2]{2}$ and $\sqrt[2]{360} = \sqrt[2]{4}\sqrt[2]{90} = 2 \cdot \sqrt[2]{90} = 2 \cdot \sqrt[2]{9} \cdot \sqrt[2]{10} = 6 \cdot \sqrt[2]{10}$.

4.3.2 Exponential notation for n-th roots

Another way to write n-th roots is as fractional exponents: $\sqrt[n]{x}$ is also written as $x^{1/n}$. This makes sense because of the rule I told you about raising exponentiated numbers to a power: the number $x^{1/n}$ should satisfy $(x^{1/n})^n = x^{(1/n) \cdot n} = x$.

4.3.3 Using exponent rules to find n-th roots

The exponent interpretation for n-th roots suggests several ways that they can be computed. Here are a few examples:

$$\sqrt[3]{a^6} = a^{(1/3) \cdot 6} = a^2.$$
$$\sqrt[6]{a^3} = a^{3/6} = \sqrt[2]{a}$$
$$\sqrt[4]{a} = a^{1/4} = a^{(1/2) \cdot (1/2)} = \sqrt[2]{\sqrt[2]{a}}$$
$$\sqrt[6]{a} = a^{1/6} = a^{(1/3) \cdot (1/2)} = \sqrt[2]{\sqrt[3]{a}} = \sqrt[3]{\sqrt[2]{a}}$$

Rules like these can be used to find some exact formulas. For example, $\sqrt[6]{27} = \sqrt[6]{3^3} = \sqrt[2]{3}$ and $\sqrt[4]{144} = \sqrt[2]{12} = 2\sqrt[2]{3}$. You should get used to doing calculations like these. They come up all the time. Here are a few more to practice on:

Question: Simplify $\sqrt[9]{8000}, \sqrt[3]{54}, \sqrt[3]{343^2}, \sqrt[4]{256}, \sqrt[5]{160}$ *and* $\sqrt[2]{14641}$.

The answers are $2\sqrt[2]{5}$, $3\sqrt[3]{2}$, 49, 4, $2\sqrt[5]{5}$ and 121.

They can also help you find approximations to roots when you need them. For example, $\sqrt[6]{900} = \sqrt[3]{30} \approx 3.1$. This is one of the few areas in the arithmetic category where you sometimes get the opportunity to find a clever solution.

4.3.4 Simplest form answers

When answering questions about fractions you're probably used to putting answers in simplest form: writing the fraction as a/b where a and b have no common factors. When they ask you to write an answer as a mixed number in simplest form, they want an answer like 5⅔: a whole number plus a simplified fraction that is between zero and one.

When answering questions about roots an answer is considered to be in *simplest radical form* only if it has several properties. (1) $\sqrt[n]{a}$ is not in simplest form if some n-th power is a nontrivial factor of a. For example, $\sqrt[4]{32}$ should be simplified to $2\sqrt[4]{2}$. (2) $\sqrt[n]{a}$ is not in simplest form if a is an m-th power and m is a factor of n. For example, $\sqrt[4]{4}$ should be simplified to $\sqrt[2]{2}$. (3) Fractions should not have roots in the denominator. The way you get rid of a root in the denominator of an expression like $1/\sqrt[2]{3}$ is to multiply the top and bottom by $\sqrt[2]{3}$. This lets you write the fraction as $\sqrt[2]{3}/3$.

4.4 IMLEM Questions

The meet #3 arithmetic category has been one of the most formulaic parts of any IMLEM meet. In four of the last six years the meets have been nearly identical. I'll call the common setup one type-1 questions, one type-2 question, and one type-3 question. In the other two years, two of the three questions have followed the formula.

4.4.1 Type-1 questions: long products you simplify by canceling common factors

Type-1 questions are a long product of terms involving exponents. For example:

Question: Evaluate the following expression. Express your answer as a mixed number in simplest form.

$$\left(\frac{9}{16}\right)^{-2} \cdot 6^3 \cdot 7^0 \cdot 2^{-5} \cdot \left(13^1 - 13^0\right) \cdot \left(2\tfrac{1}{2}\right)^3$$

These are not bad questions: it lets them test in one question whether you know what positive and negative and zero exponents are, how to raise a fractions and negative numbers to a power, and how to write a fraction as a mixed number. There is, however, little creativity involved. The **trick** is always the same: write everything out as a big product and don't do any multiplications or divisions until the very end because lots of factors will cancel from the numerator and denominator. For example, the product above is:

$$\frac{2^8}{3^4} \cdot \frac{2^3 3^3}{1} \frac{1}{2^5} \frac{12}{1} \frac{5^3}{2^3} = 2^5 5^3 = 4000.$$

A couple examples from past IMLEM meets are:

Questions: (IMLEM Meets #3, Jan. 2003 and 2005) Simplify the following expressions:

$$\left(\frac{2}{3}\right)^{-3} \cdot \left(\frac{9}{16}\right)^{-2} \cdot \left(\frac{7}{8}\right)^{0} \cdot \left(\frac{1}{4}\right)^{3}$$

$$19^0 + 7^4 \cdot 3^5 \cdot 7^{-3} \cdot 3^{-3} + 23^3 \cdot 23^{-3}$$

The answers are 1/6 and 65.

4.4.2 Type-2 questions: counting problems that require approximations

The IMLEM question writer does a tremendous job. This, however, is one place where he's been lazy and produced an exam that unreasonably favors students who've studied past exams. Three past questions from this category are:

Questions: (IMLEM Meets #3, Jan. 2002, 2005 and 2006):
   How many whole numbers are between $\sqrt[3]{22}$ and $\sqrt[3]{2002}$?
   How many whole numbers are between $\sqrt[3]{25}$ and $\sqrt[3]{2005}$?
   How many whole numbers are between $\sqrt[3]{26}$ and $\sqrt[3]{2006}$?

In other years the question has been a little different. Apart from the repetition, type-2 questions are a good way to test knowledge of the material and give you a chance to think.

Question: (IMLEM Meet #3, Jan. 2003): How many whole numbers are there between $(2/3)^{-2}$ and $(3/4)^{-4}$?

Question: (IMLEM Meet #3, Jan. 2001): How many whole numbers are there between $\sqrt[4]{100}$ and $\sqrt[2]{1000}$?

The key to these questions is to recognize that what you want to do is to figure out which integers the numbers are between, not to put them in simplified form or to approximate them to greater precision. Some standard ways to approximate numbers like this are:

1. Use a known reference point

    The approximation to $\sqrt[3]{26}$ is easy. We know that $\sqrt[3]{27} = 3$ and $\sqrt[3]{26}$ is sufficiently close to $\sqrt[3]{27}$ so that it's obvious that $\sqrt[3]{26}$ must be between 2 and 3.

2. Guess, check and update

    To approximate $\sqrt[2]{1000}$ the obvious reference point is $30^2 = 900$. Hence, $\sqrt[2]{1000}$ must be a little bigger than 30. Calculating $31^2 = 961$ makes you realize that $\sqrt[2]{1000}$ must be between 31 and 32.

    $\sqrt[3]{2006}$ can be done similarly. You know immediately that the answer must be a bigger than 10 because $10^3 = 1000$. You could thus start checking 11, 12, 13, and so on until you find the answer. The way I would have done it is to use $\sqrt[3]{2000} = 10\sqrt[3]{2}$ as a reference point. I know the square root of 2 is about 1.41 so the cube root of 2 must be smaller. Trying 13 I find $13^3 = 2197$, which is close enough to 2006 so that $\sqrt[3]{2006}$ must be between 12 and 13.

3. Start calculating until you can guess the answer

    With $\sqrt[4]{100}$ I'd start by simplifying to $\sqrt[2]{10}$ and then say that I know this is a little bigger than 3.

    With $(2/3)^{-2}$ simplifying all the way to 9/4 is easy. This is between 2 and 3.

    With $(3/4)^{-4}$ I first simplify to $(16/9)^2$ and then to 256/81. This is between 3 and 4.

After you're done simplifying you just subtract to find how many integers are between the two points. If you think about it for a little while it should be obvious that:

*   If x is between a and $a+1$ and y is between b and $b+1$ then the number of integers between x and y is $|b - a|$.

The answers to the questions I started this section with are 10, 10, 10, 1, and 28. Some facts that could come in handy are that $\sqrt[2]{2007}$ is between 44 and 45, $\sqrt[3]{2007}$ is between 12 and 13, $\sqrt[4]{2007}$ is between 6 and 7, $\sqrt[5]{2007}$ is between 4 and 5, and $\sqrt[6]{2007}$ is between 3 and 4. These facts, of course, are just about 2007. It turns out though that the same facts are true of all years up through 2024. Questions involving the year are popular in other categories too, so if you have time to waste in a math team meeting sometime you could spend it trying to figure out cool facts about whatever year it is.

4.4.3 Type-3 questions: complicated root expressions with simple answers

Most past IMLEM meets have also asked you to simplify some complicated expression that has lots of roots in it. Some examples are:

Questions: (IMLEM meets #3, Jan. 2001, 2002, and 2005) Evaluate the following expressions. Write your answers in simplest radical form.

$$\frac{\sqrt[6]{\left(\sqrt{81}\times\sqrt{64}\right)^3}+\sqrt[2]{\left(2\times\sqrt[3]{512}+\sqrt{256}\right)}}{\sqrt[5]{72}\times\sqrt{9^3}\times\sqrt[3]{8^2}}$$

$$\sqrt[4]{xy} \text{ if } x = 24 \text{ and } y = 54$$

There's not much I can tell you about these problems. Mostly, you just start with terms inside parenthesis (or the inside radicals if there are no parentheses) and evaluate everything.

As always, one tip is to keep everything factored (or even factor it more) in hopes that something will simplify or cancel. It usually does. For example, in the third problem if you multiply 24 by 54 and you may find yourself starting at $\sqrt[4]{1296}$ wondering what to do next. If you factor the terms as $24=2^3 3$ and $54=2^1 3^3$, then the product is $2^4 3^4$ and it's obvious what the fourth root is.

A second tip is that you can often make us of the formulas for simplifying radicals I mentioned earlier. For example, $\sqrt[2]{9^3} = \sqrt[2]{(3^2)^3} = \sqrt[2]{3^6} = 3^3$ and $\sqrt[6]{\left(\sqrt{81}\times\sqrt{64}\right)^3} = \sqrt[2]{\sqrt{81}\times\sqrt{64}}$.

Finally, an important third tip is that most of the problems seem to have very simple answers. For example, the answers to the above three problems are $10\sqrt{2}$, 6 and 6. If you come upon a term like $\sqrt[3]{343}$ your first thought should be that 343 is probably a perfect cube. (It is 7^3.) If you do a problem and come up with $\sqrt[6]{2^5}$ you should wonder whether you've done something wrong in simplifying the term inside the radical and whether it should have been 2^6 or 2^3 instead of 2^5. If you're completely stuck, guess 6. It's worked pretty well in the past. You can also guess one of Bigelow's lucky numbers, 54 or 2.

Category 5 – Algebra

The algebra topics for meet #3 are absolute values, inequalities in one variable, and interpreting line graphs. Most sixth and seventh graders (at least) will not have seen these things in their math classes.

5.1 Linear Equations with Absolute Values

The *absolute value* of a number is the magnitude of the number (ignoring the minus sign if there is one.) In other words, the absolute value of a positive number is the number itself and the absolute value of a negative number is what you get when you erase the minus sign. The notation for the absolute value of x is $|x|$. For example, $|4| = 4$ and $|-3| = 3$.

A typical algebra problem with an absolute value would be to solve $|x - 4| = 7$. The way to do problems like this is to divide them into two problems: $|x - 4| = 7$ if $x - 4 = 7$ or $x - 4 = -7$. You then solve both of these equations: the solution to the first is 11 and the solution to the second is -3. The answer to the problem is therefore $x = 11$ or $x = -3$.

A few other problems you could try for practice are:

Question: Solve $|3x - 4| = 5$.
Question: Find the positive difference between the two solutions to $|2x + 1| = 3$.
Question: Solve $|4x - 7| = 0$.

The answers to the first is $x = 3$ or -1/3. The answer to the second is 3. The answer to the third is $x = 7/4$.

5.2 Working with Inequalities

An *inequality* is a statement like $x > 17$ or $x \geq 17$. The first is read "x is greater than 17". The second is read "x is greater than or equal to 17."

Equations with equalities often have one or a few solutions, but equations with inequalities usually have an infinite number of solutions. For example the solution to $x \geq 17$ is all numbers that are 17 or larger.

5.2.1 Line graphs

A common way to represent the solution to an inequality is with a line graph. The picture below shows how we'd represent $x > 6$.

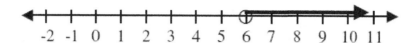

The bold ray just above the x-axis is the set of numbers that are greater than six. The line graph for $x \geq 6$ would be similar, but would start with a solid dot at six instead of an open circle. The line graph for an $x <$ or $x \leq$ inequality would point to the left. For example, the figure below shows the graph for $x \leq 5$.

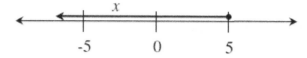

5.2.2 Solving inequalities: basics

In a typical inequality problem one is given a more complex inequality and tries to reduce it to a simple inequality like $x > 6$. For example, one could be asked to solve $2(x + 5) > 4$. Doing this is mostly just like solving equalities. The first thing you'd do in this problem is probably to multiply out the terms on the left side to give $2x + 10 > 4$. Then you'd subtract 10 from both sides to give $2x > -6$. Then you'd divide by 2 to give $x > -3$.

Here are a few more problems to try:

Question: Solve $3x - 4 < 5$.
Question: Solve $12(y+7) - 3y \geq 48$.
Question: Solve $2(z - 4) + 12 > z+1$.

The solutions are $x < 3$, $y \geq -4$, and $z > -3$.

5.2.3 Solving inequalities: multiplying by a negative number

One important thing you need to remember when doing inequality problems is that whenever you multiply or divide by a negative number you need to reverse the inequality. For example the solution to $-3x > 6$ is $x < -2$ because we divide both sides of the first equation by negative 3 to get the second. The solution to $-y/9 \leq -2$ is $y \geq 18$ because you get from the first equation to the second by multiplying both sides by negative 9.

5.2.4 Solving inequalities: IMLEM problems

Most IMLEM problems on inequalities ask you to find the value of a constant that makes a given answer correct. For example, they could ask

Question: For what value of C is the solution to $3(x+2) - 4 < 5+2C$ that $x < -1$.

The way to do a problem like this is to just solve it as you would any problem asking you to find x keeping in mind that C is a constant. In this example, a natural first step would be to multiply out the term on the left side (and then subtract the four) to reduce it to $3x + 2 < 5+2C$. Then, subtract 2 from both sides to get *3x < 3+2C*. Finally, divide both sides by 3 to give *x < 1+⅔C*. This matches the solution they gave if and only if *1+⅔C = -1*. This gives you a standard algebra problem to solve. The answer is *C = -3*.

What happened above is what will always happen. You first solve an inequality problem to get an expression for *x*. Then you solve a standard equality problem to find the constant. Often in these problems they give you a line graph instead of an inequality as the desired solution. For example,

Question: (IMLEM Meet #3, Jan. 2006) For what value of B does the solution to the inequality 3(5x-4) – 3x + B > -21 match the line graph below?

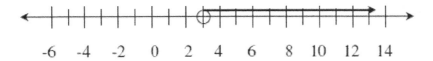

Giving you a line graph doesn't really change anything, however. The line graph tells you that the solution to the problem is supposed to be *x > 3*. Solving the problem you find $x > (-B - 9)/12$. These two match for *B = -45*.

One tip is that I find it easier in these problems to not solve the problem all the way. For example in the problem above I would have stopped at $12x > -B - 9$ and then said "OK. For the solution to this to be *x > 3* it must be that $-B - 9 = 36$."

Another tip that you can use if you're short on time is that if you know the problem has a solution (which is implied by the wording of the IMLEM question above), then you don't really have to work with inequalities at all. For example, in this problem for the line graph to be the solution to the inequality problem, the equality problem *3(5x-4) – 3x + B = -21* must have *x=3* as a solution.

A couple more practice problems along these lines are:

Question: For what value of B is the solution to 16(x-1) – 4(x – 2) < B that x < 2.
Question: For what value of A is the solution to 5(x-2) +20 > 3x+A that x > -5.

The solutions to these are 16 and 0.

5.2.5 Advanced topic: multiplying through by x

In some problems you'll find yourself wanting to multiply through by *x*. This causes a complication: you don't know when you do this whether x is positive or negative so you don't know whether you have to reverse the inequality. For example, consider

Question: For what values of x is 6/x > 3?

What you do in this situation is to break the problem up into the two cases: the solution is the set of x such that ($x > 0$ and $6 > 3x$) or ($x < 0$ and $6 < 3x$). The solution to the first subproblem is $0 < x < 2$. The second subproblem has no solutions. Hence, the answer to the whole question is $0 < x < 2$.

If IMLEM wanted to modify their standard question so that this came up the could ask:

   Question: For what value of C is the solution to $Cx - 5 < 2$ that $x > -1$.

The answer to this one is -7. The way to follow the general procedure in this problem is to divide into cases when you divide both sides by C. You could also probably do it more intuitively buy just recognizing that the only time in the process of solving the problem when the inequality could have turned around is when you divided both sides by C. Hence, C must be a negative number.

5.3 Working with Absolute Values and Inequalities

The final thing that IMLEM likes to ask you to do is to solve problems with inequalities and absolute values.

5.3.1 Absolute value as a distance measure

You can think about $|x - a|$ as measuring the distance between x and a on a number line. For example, $|x - 3| = 5$ has two solutions, -2 and 8. These are the two points that are five units away from 3 on a number line. Negative two is five units to the left and eight is five units to the right.

Some simple questions that are easy to answer this way are:

   Question: For what values of x is $|x - 17| < 4$?
   Question: How many integers n satisfy $|n - 17| \leq 5$?

The answer to the first is $13 < x < 21$. Even without finding all the values of n that solve the second we know the answer is 11: the five numbers to the left of seventeen, seventeen itself, and the five numbers to the right of seventeen.

You can similarly think about $|x + 2|$ as giving the distance between x and negative two (because $|x + 2| = |x - (-2)|$). For example, the set of solutions to $|x + 17| < 10$ is the set of all x with $-27 < x < -7$.

One slightly harder question of this variety is:

   Question: For how many integers n is $|3n - 23| \leq 2$.

The answer is 2. For n to be a solution to this equation $3n$ must be one of $\{21, 22, 23, 24, 25\}$. Two members of this set are multiples of 3. Another way to do this would be to divide by 3 to get $|n - 7\frac{2}{3}| \leq \frac{2}{3}$. There are two integers with at a distance of two-thirds or less from $7\frac{2}{3}$: 7 and 8.

Another question they could ask is:

Question: For how many integers n is 15/|n-2| an integer.

The answer is eight. $15/|n-2|$ is an integer if and only if $|n-2|$ is a factor of 15, i.e. if it is 1, 3, 5 or 15. This means that any integer which is at a distance of 1, 3, 5, or 15 from 2 has the desired property. Clearly there are eight such integers: the integers that are 1, 3, 5, and 15 units to the left of two and the integers that are 1, 3, 5, and 15 units to the right of 2.

5.3.2 A quick caution: Don't multiply by zero

In section 5.2 I explained that you could multiply both sides of an inequality by a positive number or change the direction of the inequality and multiply both sides by a negative number. One thing you need to be careful not to do with a strict inequality like $x < 3$ is to multiply both sides by zero. If you did, you'd end up with $0 < 0$, which is false.

5.3.3 IMLEM Problems

Most IMLEM problems involving both absolute values and inequalities have been things like:

Question: (IMLEM Meet #3, Jan. 2002) How many integer values of n satisfy |15/n| > 4?

Question: (IMLEM Meet #3, Jan. 2006) How many integer values of n satisfy |18/(n+1)| > 3?

The first step in the first problem like this is to recognize that $|15/n| = 15/|n|$. Hence, if $n \neq 0$ we can multiply both sides by the positive number $|n|$ and find that the equation is true if and only if $15 > 4|n|$ or $|n| < 15/4 = 3.75$. The answer is six: n must be one of -3, -2, -1, 1, 2, or 3. I made the caution a separate section above because I'm sure most people who got this problem wrong forgot that zero was not a valid solution and said seven. Zero is not a solution because 15/0 is not well defined.

The answer to the second problem is 10. Again we can multiply through by $|n+1|$ if $n \neq -1$ to find the equation holds when $18 > 3|n+1|$ or $|n+1| < 6$. There are ten solutions to this (after ruling out $n = -1$): the five integers to the left of -1 and the five integers to the right of -1.

Here's two final problems for practice:

Question: How many integer values of n satisfy |28/(n+14)| ≥ 8?
Question: How many integer values of n satisfy |15/(3n+1)| > 5?

The first problem works just like the ones above and the answer is 6. The second is a little different. The answer is 2.

A final re-reminder: please be careful not to count the potential solution that is not a solution because it would involve dividing by zero.

IMLEM Meet #4

Meet #4 is the first calculator meet. Be sure to bring one because you'll really need it–they ask lots of questions where you need to do things like multiply 144 by 3.14 or compute 1.004^{12}!

In terms of content, most categories are a mix of some topics you've probably seen and some topics you won't have seen unless you do math team.

Category 2 – Geometry

Most geometry problems in meet #4 only require that you know a small number of facts about circles. For most of them you just need to know what's on the first and fifth pages of this chapter. Most of you already know what's on page 1, so if you have very little time, read page 5. Some problems require that you haven't forgotten your triangle and polygon facts, so I include a quick review. I also include a lengthy section of advanced facts that are fun and could come up even though year after year after year they never seem to.

2.1 Areas and Perimeters of Circles

Most of you have probably had some class that taught about circles. The basic definitions are:

- A *circle* is the set of all points equidistant from one point, which is called the center of the circle.

- The *radius r* of a circle is the distance from the center to all the points.

- A *diameter* is a line segment that passes through the center of the circle and has both endpoints on the circle. The length D of any diameter is twice the radius: $D=2r$.

- The *circumference* of a circle is the distance around the outside of the circle (this is like the perimeter).

- The *area* of a circle is the number of square units of space contained within it.

2.1.1 The important formulas

The two really important formulas about circles are:

- The circumference of a circle is $C=2\pi r= \pi D$.

- The area of a circle is $A = \pi r^2$.

IMLEM wouldn't ask you to just plug numbers into these formulas, but sometimes they are a lot like asking you to plug in a couple times. For example,

Question (IMLEM Meet #4, Feb. 2002): A circle has a circumference of 18.84 centimeters. How many square centimeters are in the area of the circle. Use π = 3.14 and express your result to the nearest hundredth of a square centimeter.

A good way to do this problem is to first use the circumference formula to find the radius: $2\pi r = 18.84$ implies $r = 18.84 / (2\pi) \approx 3$. Then you use the area formula: $A = \pi 3^2 \approx 28.26$. Be sure to note IMLEM's rules for approximating: if they ask you to use 3.14 for π this is what you need to do. It's not OK to use a different or better approximation to π or to do any intermediate rounding off.

My two step approach to the problem above didn't really need any algebra. Sometimes, however, a little algebra is really necessary.

Question (IMLEM Meet #4, Feb. 2004): Find the number of feet in the radius of a circle whose area given in square yards is numerically equivalent to its circumference given in feet.

At first, this could sound hard. If you use a little algebra, however, it becomes easy. One square yard is nine square feet (draw a 3 x 3 grid of boxes if this is not obvious). Hence, what the problem is telling you is that $\pi r^2/9 = 2\pi r$. The solution to this is $r = 18$.

2.1.2 Areas of other figures

Areas of lots of other figures can be simply derived from the ones above. For example, the area of a semicircle is ½ and the area of a quarter of a circle is ¼ of the area of a circle.

The areas of lots of other figures can be found by addition or subtraction. Here are a couple simple examples:

Question: The figure on the left below shows two circles. The larger has a radius of R. The smaller has a radius of r. Find the area of the shaded region.

Question (IMLEM Meet #4, Feb. 2004): A circle is inscribed in a square whose area is 2.25 square inches. How many square inches are there in the sum of the areas of the two shaded regions.

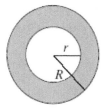
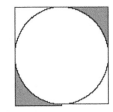

The area of the ring is easily found by subtraction. The area of the larger circle is πR^2. Subtracting off the area of the smaller circle, πr^2, gives that the area of the ring is $\pi (R^2 - r^2)$.

To do the second problem you first note that the fact that the square has area 2.25 implies that its side-length is 1.5. If the square has side s, then the circle has radius $s/2$. Hence the circle has radius 0.75. If they'd shaded all four corner regions, you'd do the problem by subtracting the area of the circle from the area of the square and get $2.25 - \pi (0.75)^2$. They only asked for the area of two of the four regions, so the answer is half of this: $\frac{1}{2} (2.25 - \pi (0.75)^2)$.

In more complicated figures you may need to both add and subtract. Once a problem gets complicated there are usually several ways to get the answer. It's usually a good idea not to worry about this. Just pick one method and use it.

Question (IMLEM Meet #4, Feb. 2006): The side length of the square in the figure below is 2 inches. The arcs are all 90 degree arcs of circles with radius 1 inch and the small white circle has a diameter of 1 inch. How many square inches are in the area of the shaded region?

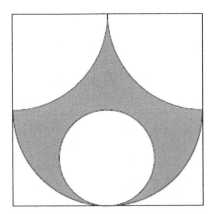

One way to do this would be to first add up the areas of the five white regions. The upper left and right regions are quarter circles. You can find the areas of the bottom left and bottom right regions by subtracting the area of a quarter circle from the area of a quarter square. Finally, you add the area of the circle in the bottom. You'd then get the area of the shaded region by subtracting the sum of these five from the area of the square.

There's also a neater solution to the problem. If you stare at the figure you'll notice that the white regions in the bottom left and bottom right corners could be rotated up to fill in the shaded parts of the top half. If you did this, the area of the shaded part would just be the area of the bottom half of the rectangle, 1×2, minus the area of the circle on the bottom, $\pi \left(\frac{1}{2}\right)^2$.

2.2 Arcs and Angles

There are some interesting theorems about angles related to circles that tell you things you wouldn't have guessed that are hard to work out.

2.2.1 Definitions

- A *central angle* is an angle with its vertex at the center of the circle and its endpoints on the circle. In the figure on the left below below AOC is a central angle.

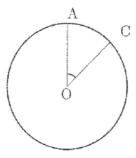
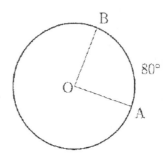

- An *arc* is the part of the circle between two points. In the figure on the right above, arc AB is the part of the circle between A and B. You'll sometimes here people referring to the 80 degree arc in the figure above as *minor arc* AB. You could also get from A to B by going around the circle the long way. This 280 degree arc would be called *major arc* AB. If someone doesn't specify which one they mean you should always assume it's the minor arc.

- The *arc angle* is a measure of the relative size of an arc. Arc angle AB is defined to be the measure of central angle AOB. In the figure on the right above, *x=80*.

- An *inscribed angle* is an angle that has its vertex on the circle and both of its endpoints on the circle. In the figure on the left below ABC is an inscribed angle. In the figure on the right, ADB is an inscribed angle.

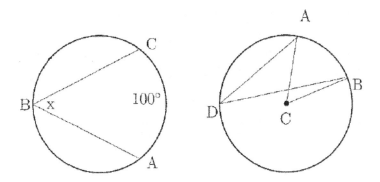

2.2.2 The important facts

The most important thing to know about angles in circles is:

- The measure of an inscribed angle is one-half of the measure of the arc it intersects. In the figure on the left above, the measure of ABC is 50°. In the figure on the right above the measure of ADB is one half of the measure of ACB.

A couple of problems in which this lets you get answers very fast are:

Question (IMLEM Meet #4, Feb. 2003): In the figure on the left below the measure of minor arc AB is 70 degrees. How many degrees are in the measure of the inscribed angle ADB?

Question (IMLEM Meet #4, Feb. 2004): Segment AC is a diameter of the circle at right below. If the measure of angle CAB is 49 degrees, how many degrees are in the measure of angle ACB?

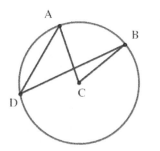
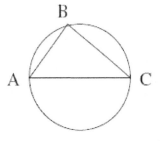

The first one is an immediate application of the Inscribed Angle Theorem. The measure of ADB is one-half of the measure of arc AB. The answer is 35 degrees.

One way to do the second problem is subtracting arcs: arc AC is 180 degrees. Because angle CAB is an inscribed angle for arc BC, we know that arc BC has measure 49 × 2 = 98 degrees. By subtraction arc AB has measure 180 – 98 = 82 degreees. ACB is inscribed in this arc, so its measure is 41 degrees.

My favorite way of doing the second problem is different. Draw in a point O at the center of the circle. (This is in the middle of segment AC.) AOC is a straight line, which means that it's a 180 degree angle. Hence, angle ABC is one-half of this or 90 degrees. You then get angle ACB using the fact that the angles in a triangle add up to 180 degrees: ACB = 180 − (90 + 49) = 41 degrees.

The idea of putting that point in the middle of the figure might escape you during a meet so I thought I'd put in the main implication with a bullet point.

- An inscribed angle that intersects the endpoints of a diameter is a right angle.

One other fact that comes in handy is not really a circle fact, but just a simple implication of the fact that all radii of a circle have the same length.

- Any triangle formed by connecting the center with two points on the circle is an isosceles triangle. This implies that the angles opposite the two radii are equal.

An example of a problem where this lets you get an answer very fast is:

Question (IMLEM Meet #4, Feb. 2005): In the figure below, point B is the center of the circle. The measure of ACB is 65 degrees and the measure of angle BCD is 14 degrees. How many degrees are in the measure of angle ADC?

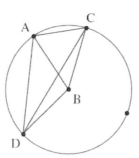

The easy solution is to first use the isosceles triangle to say that BAC is also 65 degrees. By subtraction, ABC is 50 degrees, which implies that minor arc AC is also 50 degrees. ADC is an inscribed angle in this arc, so its measure is 25 degrees. This solution is not only quick, but illustrates that the fact that BCD was 14 degrees was a completely unnecessary for doing this problem.

One final fact I thought I'd mention in this section is:

- The area of the sector of a circle bounded by a central angle of measure $x°$ is $(x/360) \pi r^2$.

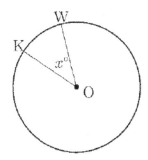

For eighth graders who are getting a little bored seeing the material for the third time, a good challenge problem is to try to prove that the measure of an inscribed angle is half of the measure of the central angle. The general case is very hard. A special case that has a nice proof is where the line from the vertex of the inscribed angle through the center of the circle bisects the inscribed angle.

2.3 A Quick Review of Triangles and Polygons

Here are a few things you might want to review.

2.3.1 Angles

Some facts about angles in polygons are:

- The angles in a triangle add up to 180°.

- The angles in a quadrilateral add up to 360°.

- The angles in an n-gon add up to (n-2) 180°.

- The exterior angles in an n-gon add up to 360°.

- Angles opposite equal sides in an isosceles triangle are equal.

2.3.2 Area

Three formulas for computing the area of a triangle are:

- $A = \frac{1}{2} b\, h$, where b is the base of the triangle and h is the height.

- $A = \sqrt{s(s-a)(s-b)(s-c)}$, where a, b, and c, are the sides and s is one-half of the perimeter.

- If one vertex of a triangle is at (0,0) and the other two at (x_1, y_1) and (x_2, y_2), then the area is $A = \frac{1}{2} | x_1 y_2 - x_2 y_1 |$

2.4 Advanced Circle Facts

I haven't seen any of these appear on IMLEM meets, but there are a lot of other interesting facts about circles. This section is already pretty long so you should feel even more free than ever to skip reading the rest, but some of the results are cool so I put them in.

2.4.1 Tangents

A *tangent* is a line that intersects a circle at exactly one point.

Two facts about tangents are:

- There is a unique tangent through any point on a circle.

- A tangent line is perpendicular to the line connecting the center of the circle to the point of tangency.

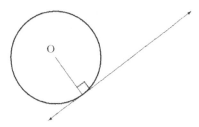

2.4.2 Inscribed circles

An *inscribed circle* is a circle that is tangent to all three sides of a triangle.

Three facts about inscribed circles are:

- Every triangle has a unique inscribed circle.

- The center of the inscribed circle is located at the intersection of the angle bisectors of the three angles of a triangle.

- A formula for the area of a triangle is $K = i\, s$, where i is the radius of the inscribed circle and s is the semiperimeter (which is just a fancy word for half of the perimeter). You can see this by adding up the subtriangles in the figure below.

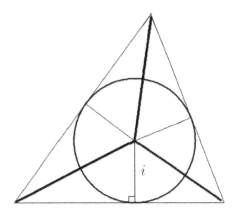

2.4.3 Circumscribed circles

A *circumscribed circle* is a circle that passes through all three vertices of a triangle.

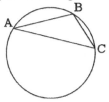

Four facts about circumscribed circles are:

- Every triangle has a unique circumscribed circle.

- The center of the circumscribed circle is at the intersection of the perpendicular bisectors of the three sides.

- A formula for the area of a triangle is A = $a\,b\,c\,/\,(4R)$, where a, b, and c are the lengths of the sides and R is the radius of the circumscribed circle.

- In a right triangle the center of the circumscribed circle is the midpoint of the hypotenuse. R = ½ c

2.4.4 Cyclic quadrilaterals

Every triangle has a unique circumscribed circle. This is not true of quadrilaterals. Most don't have a circumscribed circle. To see this, pick any three vertices. The circumscribed circle for that triangle is the unique circle that passes through these three points. If it doesn't happen to also pass through the fourth vertex, then the quadrilateral can't have a circumscribed circle.

If a quadrilateral has a circumscribed circle, then it's called a *cyclic quadrilateral.*

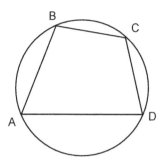

There are three main facts about cyclic quadrilaterals:

- Opposite angles are complements, i.e. they add up to 180°.

- The area of a cyclic quadrilateral is given by A = $\sqrt{(s-a)(s-b)(s-c)(s-d)}$

- If e and f are the lengths of the diagonals in a cyclic quadrilateral, a and c are the lengths of one pair of opposite sides, and b and d are the lengths of the other pair of opposite sides, then $ef = ac + bd$.

The second fact is a generalization of the formula for the area of a triangle. Think about a cyclic quadrilateral where two vertices are very close together so the length of side d is approximately zero.

The third fact is called Ptolemy's theorem. Note that it is closely related to the Pythagorean theorem. Any rectangle is a cyclic quadrilateral. If the short side of the rectangle has length a, the long side has length b, and the diagonal is c, then Ptolemy's theorem says $c^2 = a^2 + b^2$.

The converse of Ptolemy's theorem is also true: if $ef = ac + bd$, then the quadrilateral can be inscribed in a circle.

One other fact about cyclic quadrilateral is

- The product of the distances from the diagonal intersection to the circle is the same on each diagonal. In the figure above, for example, if we call the intersection of the diagonals E then we have EA × EC = EB × ED.

2.4.5 More angle facts

Some formulas related to the inscribed angle formula are:

- If two chords intersect inside a circle, then the measure of the angle between them is one-half of the sum of the measures of the two intersected arcs. In the figure below $x = \frac{1}{2}(170 + 70) = 120°$.

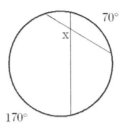

- If the vertex of an angle is outside of the circle then the measure of the angle is one-half of the difference between the two incepted arc. In the figure below, $x = \frac{1}{2}(80 - 20) = 30°$.

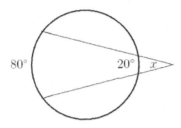

- The measure of a tangent angle is also one-half of the measure of the arc it intersects. In the figure below, $x = \frac{1}{2}(120) = 60°$.

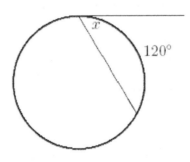

Category 3 – Number Theory

The number theory category in meet #4 covers two unrelated topics: sequences and series; and modular arithmetic. In 2001 it was a really hard category. In the last couple years it hasn't been so bad. If you only have time to read one page, read the first page of these notes: the material they cover comes up all the time. There are also always questions on modular arithmetic. It's my favorite part of meet #4: there's lots of new math on it and once you learn it you can answer questions that sound amazingly hard. It can take a little while to get used to, however.

3.1 Basic Sequences and Series

3.1.1 Sequences

- An *arithmetic sequence* is a sequence of numbers such that the difference between adjacent terms is equal. For example, 5, 7, 9, 11, 13 ... and 13, 10, 7, 4, 1, -2, -5 ... are arithmetic sequences. 1, 4, 9, 16, 25 ... and 1, 2, 4, 8, 16, 32 ... are not.

- If $a_1, a_2, a_3, \ldots$ is an arithmetic sequence then the nth term is $a_n = a_1 + (a_2 - a_1)(n-1)$.

This formula is useful for answering questions like:

Question (IMLEM Meet #4, March 2003): Find the 48th term of the following sequence: 23, 32, 41, 50, ...

Question: Find the 26th term of the following sequence: 18, 12, 6, 0, -6, ...

The answers to these are 446 and -132.

3.1.2 Series

A series is a sequence of numbers that are added together. I'll write S_n for the sum of the first n terms of the series: $S_n = a_1 + a_2 + a_3 + \ldots + a_n$.

There are several formulas for the sum of an arithmetic sequence. Suppose $a_1, a_2, a_3, \ldots$ is an arithmetic sequence and $S_n = a_1 + a_2 + a_3 + \ldots + a_n$. Then,

- $S_n = (n/2)(a_1 + a_n)$.

- $S_n = n\, a_1 + \frac{1}{2}(n-1)\, n\, (a_2 - a_1)$.

- $S_n = n\, \text{Median}\{a_1, a_2, a_3, \ldots, a_n\}$.

The first formula is easier to remember and I find it easier when the series is very short. When a series is longer you can either memorize the second formula or take a two-step approach: first find the last term; then use the first formula.

Here are some practice problems:

Question: What is the sum of the first 33 terms of the following sequence: -7, 4, 15, 26, ...?
Question: Find the sum of the first 20 positive odd integers.

Question (IMLEM Meet #4, Feb. 2005): The measures in degrees of the three angles of a triangle form an arithmetic sequence. How many degrees are in the measure of the second largest angle?

The answers are 5577, 400, and 60. The last problem is an example of one where the third formula comes in handy.

3.2 Modular Arithmetic

3.2.1 Basic facts

In modulo N arithmetic (also called mod N) the numbers are 0, 1, 2, 3, ..., N-1, 0, 1, ..., N-1, 0, 1, 2, ... People sometimes picture this by thinking of numbers as lying on a number circle instead of on a number line.

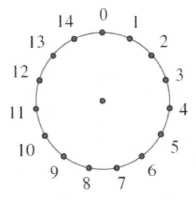

An alternate way to think about modular arithmetic is in terms of remainders:

- If r is the remainder when y is divided by x, then $y=r$ (mod x).

All of the formulas you use for adding, subtracting, and multiplying regular numbers still work in modulo N arithmetic. The only thing to watch out for is that numbers have multiple names. For example, -2=13 (mod 15) and 100=4 (mod 8). Examples of things you can still do in modular arithmetic are:

- $a + b = b + a \pmod{N}$

- $a\,(b + c) = ab + ac \pmod{N}$

- $(ab)\,c = a\,(bc) \pmod{N}$

- $a^{bc} = (a^b)^c \pmod{N}$

3.2.2 Simplifying expressions

Often, problems that sound very complicated are not very complicated if you remember that mod N numbers can be simplified. For example, if you are asked to simplify 63 (55 + 16) modulo 10, you just write that 63 (55 + 16) = 3 (5 + 6) (mod 10) = 3 × 1 (mod 10) = 3 (mod 10).

Here are some practice problems:

Question: Simplify 24 + (6 × 24) (mod 11).

Question (IMLEM Meet #4, Feb. 2003): Amanda bought a new remote control for her television. It allows her to advance the channel number by fives. For example if she's watching channel 12 and hits the "advance five" button, the television switches to channel 17. Amanda has exactly 43 channels numbered 1 to 43 and the next channel after 43 would be channel 1. If she starts out on channel 12 and presses the "advance five" button 27 times in a row, what channel will she end up watching?

Question (IMLEM Meet #4, Feb. 2004): Coach McMod gave his fifteen basketball players numbered 0 to 14. For passing practice the players form a circle so that their jersey numbers go in clockwise ascending order. Last practice, Coach McMod gave the ball to the player with jersey number 6 and told the players to pass to the fourth person on their left. After 71 passes what was the number of the player holding the ball.

The first one can be simplified as 24 + (6 × 24) (mod 11) = 2 + (6 × 2) (mod 11) = 3 (mod 11). You'll notice from the other two that IMLEM likes to make up contrived word problems that you do with modular arithmetic. The first word problem above is asking you to simplify 12 + (27 × 5) (mod 43). The answer to this is 18. The second is asking you to simplify 6 + (71 × 4) (mod 15). (Be sure you understand why its mod 15 and not mod 14 and why you're adding 4 instead of subtracting 4 with each pass.) The answer to this is 5.

To compute powers of numbers in modular arithmetic, you first simplify the number that is being raised to a power and then multiply it by itself over and over. This is not as hard as you might think because you can simplify as you go and use the $a^{bc} = (a^b)^c$ formula. For example, if asked to simplify $13^8 \pmod{8}$, start by saying that because 13=5 (mod 8) this is the same as 5^8 (mod 8). $5^2=25=1$ (mod 8), so you know that $5^8=(5^2)^4=1^4$ (mod 8)=1 (mod 8).

Here's another practice problem that doesn't work out neatly:

Question: Find the value of 2^{35} in modulo 29.

One way to start on this one is to use $2^{35} = (2^5)^7 = 3^7$ (mod 29). Then, remember that you have your calculator with you and that you can compute 3^7 on it, divide by 29, subtract off the integer part, and multiply by 29 to figure out what the remainder is. (Some calculators have an integer-divide-with-remainder button that makes this even easier.) The answer is 12

3.2.3 Two things you can't do

Modular arithmetic works so much like regular arithmetic that you may start to feel after fifteen minutes of practice that you have the topic down cold and can simplify or solve anything using the same techniques you've used for years in regular arithmetic. Be careful. The list of things I told you you could do in section 3.2.1 are NOT all of the things you're used to doing in arithmetic. This is not just because I'm lazy. Some other things don't work. In this section I point out two places where people new to modular arithmetic sometimes make mistakes.

The first important example of something you *can't* do in modular arithmetic is simplify equations by dividing both sides by the same number.

- Example: $2x = 6$ (mod 8) does not imply that $x=3$ (mod 8). A second solution is $x=7$!

A second thing you *can't* do in modular arithmetic is to simplify numbers in the exponent.

- Example: $2^{11} \neq 2^1$ (mod 10)

You can't simplify in this way because the eleven in the exponent is NOT a modulo ten number. It's a regular number. Exponentiation is a shorthand notation that tells you you're supposed to multiply eleven 2's together to get the answer. Changing this number would be changing the question from "Evaluate $2 \times 2 \times 2 \times 2 \times 2 \times 2 \times 2 \times 2 \times 2 \times 2 \times 2$ in modulo 10" to "Evaluate 2 in modulo 10". These are unrelated questions that have different answers.

(Section 3.2.6 will discuss a special situation in which you can do something like dividing and something like simplifying exponents).

3.2.4 Solving equations

Sometimes you may be asked to solve equations in modular arithmetic. For example,

Question: Find the whole number value of x such that $0 \leq x < 11$ and $4x = 1$ (mod 11).

Question (IMLEM Meet #4, Feb. 2005): List all the solutions to the equation $6x + 5 = 14$ in Modulo 15, where x is limited to the whole numbers from 0 to 14.

There are a few ways to solve equations like this. In this section I'll just discuss <u>method #1</u>: simplify things that are easy to simplify and then just try all the numbers.

The trying out numbers method is made possible by a nice features of modular arithmetic: there are only 11 numbers to try in Mod 11 instead of infinitely many. Another thing to note is that you don't have to try all the numbers by plugging in each one and seeing if it works. Instead, you can just count by 4's and try all numbers this way. The way to count by 4's in mod 11 is: 4, 8, 12=1, 5, 9, 13=2, 6, 10, 14=3, 7, 0. The unique solution is 3.

In the second problem, you can first subtract 5 from both sides to get $6x = 9$ (mod 15) and then start counting by sixes. The answer is that 4, 9, and 14 are all solutions.

One thing that's useful to note is that the number of solutions that an equation has is different in modular arithmetic. In traditional arithmetic a linear equation will always have one solution. In modular arithmetic linear equations can have zero, one, or several solutions. It's not so hard to understand why.

The equation $4x = 1$ (mod 11) has a unique solution (for x between 0 and 10) because if there was a second number y that also satisfied $4y = 1$ (mod 11) we could subtract the two equations from each other and find $4(x - y) = 0$ (mod 11). Because 11 is prime, this implies that $x - y$ is a multiple of 11.

An equation of the form $6x = a$ (mod 15) will always have either zero or three solutions. Because $6 \times 5 = 0$ (mod 15) if you find one value of x that satisfies the equation you can find two more by adding 5 and 10 to your initial solution.

A final practice problem is:

Question: List all the solutions to the equation $6x = 12$ in Modulo 14, where x is limited to the whole numbers from 0 to 13.

The answers are 2 and 9. Do you see how you could have gotten this right away?

3.2.5 Problem solving tip: it's sometimes easier to use negative numbers

Another thing to keep in mind is that negative numbers work perfectly fine in modular arithmetic and can often simplify problems. Here are two examples:

Question: Find all values of x for which $11x = 7$ modulo 12.

Because $11 = -1$ (mod 12) this is the same as asking for solutions to $-x = 7$ modulo 12. Obviously, the answer to this is $x = -7$ (mod 12). You should always write your answer in standard form though. Here that's $x = 5$ (mod 12).

Question: Simplify 31^{169} in modulo 16.

This one sounds hard, but is also easy if you realize $31 = -1$ (mod 16). Hence, $31^{169} = -1^{169}$ (mod 16) = -1 (mod 16) = 15 (mod 16).

3.2.6 Mod p arithmetic when p is prime.

When p is a prime number modulo p arithmetic works even better. This is all because of one famous theorem about modulo p arithmetic (known as Fermat's little theorem).

- If p is prime and a is not a multiple of p then $a^{p-1} = 1 \pmod{p}$.

One implication of this is that you can do the something like (but not exactly like) one thing I said you couldn't do: simplify numbers in exponents. The not exactly like part is that **exponents are simplified using mod p-1 arithmetic instead of mod p arithmetic**. For example, we can very quickly compute $2^{53} = 2^3 \pmod{11} = 8 \pmod{11}$.

Another implication is that every number (other than zero) has an inverse in modulo p arithmetic. The inverse of a is the number b such that $ab = 1 \pmod{p}$. In regular arithmetic we often write a^{-1} for the inverse of a. There are two nice properties of inverses in mod p arithmetic.

- Every nonzero number has exactly one inverse.

- One can find the inverse of a using $a^{-1} = a^{p-2} \pmod{p}$.

Here's a sample question that you can do in two ways using these two facts:

Question: What is the inverse of 8 in modulo 17 arithmetic.

The answer is 15. One way to get this is to say that the inverse of 8 in modulo17 arithmetic is 8^{15}. This might sound hard to compute, but it's not. $8^{15} = (2^3)^{15} \pmod{17} = 2^{45} \pmod{17} = 2^{13} \pmod{17}$. We know $2^4 = 16 = -1 \pmod{17}$. Hence, $2^{13} = 2^4 \times 2^4 \times 2^4 \times 2 = -2 \pmod{17} = 15 \pmod{17}$. You could also have done this by exploiting your calculator halfway through. A calculator can't compute 8^{15} accurately, but it can compute 2^{13}, so once you got here you could just compute $2^{13} = 8192$, divide this by 17, subtract off the integer part of the answer, and multiply by 17.

A second way to have approached this would have been to say that you know 8 has an inverse mod 17, so you can just start trying numbers in order, i.e. compute 8×1, then 8×2, and so on… until you find an answer. This will often be faster. In this case, because the answer is 15, it would probably have been slower.

Here's an example of an IMLEM problem that asks you to find an inverse:

Question (IMLEM Meet #4, Feb. 2003): In modular arithmetic, a "unit" is a number that has one as a multiple in a particular modular system. Let's say a "unit pair" is a pair of numbers whose product is one and a "unit partner" is the other number in a unit pair if one of the numbers is known. For example, in Mod 7 the unit partner of 2 is 4. What is the unit partner for 8 in Mod 19?

The answer is 12. One good way to do this is to use the fact that $2^{18} = 1$ (mod 19) to say that $2^3 \times 2^{15} = 1$ (mod 19), so you can find the answer by computing 2^{15}. Another way, of course, is just to count by 8's until you get to 1: 8, 16, 5, 13, 2, 10, 18, 7, 15, 4, 12, 1. The counting way requires less thought and is probably a faster way to do this one.

Finding inverses is the key idea behind method #2 for solving equations. Consider,

Question: Solve 8x + 3 = 7 (mod 19).

As before, the best way to start is to subtract 3 from both sides to get $8x = 4$ (mod 19). Now that you know about inverses, however, you can take a clever second step: multiply both sides by the inverse of 8 in mod 17. From doing the previous problem we know that this is 12. Multiplying both sides of the above equation by 12 gives $8 \times 12 \times x = 48$ (mod 19), which is the same as $x = 10$ (mod 19).

Remember, you can only simplify exponents like this and find inverses in modulo p arithmetic if p is prime. In modulo 8 arithmetic, for example, 2 doesn't have an inverse.

3.3 Advanced Sequences and Series

Again, these advanced facts haven't really come up in past IMLEM meets (except for one question which was about a quadratic sequence.) Many are more interesting than the facts that do come up, however, so I thought I should include them.

3.3.1 Polynomial Sequences

A polynomial sequence is a sequence in which the nth term is a polynomial function of n. Some examples of this are:

- Constant sequences like 17, 17, 17, …. Here, $a_n=17$.

- Arithmetic sequences like 1, 4, 7, 10, … There is always a formula for the nth term of the form $a_n = a + b\,n$.

- Quadratic sequences like 1, 4, 9, 16, 25 … and 2, 5, 10, 17, 26, …. In these, the formula for the nth term is of the form $a_n = a + b\,n + cn^2$. The second example above is $a=1$, $b=0$, and $c=1$.

- Cubic sequences like 0, 6, 24, 60, 120, 210, 336 … . If you couldn't figure out the pattern it's $a_n = -n + n^3$.

I only know a few facts about polynomial sequences:

- You can figure out the next term in any polynomial sequence by writing the difference between every pair of adjacent terms below them, then writing down the differences of that sequence, then writing down the differences of that sequence, etc. until you get a constant sequence. For example, given the sequence 2, 5, 10, 17, 26, ... you write down the differences as shown on the left.

2 5 10 17 26		2 5 10 17 26 37 ___ ___ ___
3 5 7 9	$\rightarrow$	3 5 7 9 11 13 15 17 ___
2 2 2		2, 2 2 2 2 2 2

Seeing that the last row is all 2's you can extend the original sequence by writing down many more 2's then filling in the middle, and then the top sequence. I'll leave it to you to fill in the rest.

IMLEM did once ask about a quadratic sequence:

Question (IMLEM Meet #4, March 2001): If 17 is the first term of the sequence shown below, find the value of the 8th term: 17, 20, 26, 35, 47, ...

The answer to this one is 101. The easiest way to do it is probably to fill in a chart of differences like I did above. You could also use the formula below. The formula is a better technique if they ask you for something like the 85th term, but it comes up sufficiently rarely (both in math contests and real life) that it's probably not worth memorizing.

- A formula for the nth term of a quadratic sequence is

$$a_n = a_2 + \tfrac{1}{2}(a_3 - a_1)(n-2) + \tfrac{1}{2}(a_3 - 2a_2 + a_1)(n-2)^2.$$

- If the nth term of a sequence is a kth degree polynomial, then the nth term of its difference sequence is a $(k\text{-}1)$st degree polynomial. The difference sequence of a quadratic sequence is an arithmetic sequence. If you start with a sequence where the nth term is a kth degree polynomial, then the $(k+1)$st row is constant.

3.3.2 Geometric Sequences and Series

A geometric sequence is a sequence in which the ratio of each pair of adjacent terms is the same. For example, 1, 2, 4, 8, 16, ... and 4, 6, 9, 13½, 20¼, ... are geometric sequences. These come up all the time in math (and in MathCounts), but don't seem to come up often in IMLEM.

- The nth term of a geometric series is $a_n = a_1 (a_2 / a_1)^{n-1}$.

- A formula for the sum of the first n terms of a geometric series is

$$S_n = a_1(1 - (a_2 / a_1)^n) / (1 - (a_2 / a_1)).$$

To understand why this works think about multiplying out $(1 + r + r^2 + \ldots + r^{n-1})(1 - r)$

- A formula for the sum of an infinite geometric series is $S_n = a_1 / (1 - (a_2/a_1))$. One example of this is that the repeating decimal $0.999999\ldots = 1$. The repeating decimal is $(9/10)+(9/100)+(9/1000)+\ldots$ This is a geometric series with $a_1=9/10$ and $a_2/a_1=1/10$.

3.3.3 Polynomial Series

There are few facts about polynomial series:

- $1 + 2 + 3 + 4 + \ldots + n = \frac{1}{2} n (n + 1)$

- $1 + 2^2 + 3^2 + 4^2 + \ldots + n^2 = n (n + 1) (2n + 1)/6.$

- $1 + 2^3 + 3^3 + 4^3 + \ldots + n^3 = (1 + 2 + 3 + 4 + \ldots + n)^2 = (\frac{1}{2} n (n + 1))^2$

- If a_n is a kth degree polynomial of n then the formula for S_n is a $(k+1)$st degree polynomial in n.

- A fact that's not really about polynomial series, but which I wanted to put in anyway is that

$$1 + 1/2^2 + 1/3^2 + 1/4^2 + 1/5^2 + \ldots = \pi^2 / 6.$$

Category 4 – Arithmetic

The arithmetic category in meet #4 covers two topics: percentages and compound interest. I think of it as usually being the easiest category of the meet, although this is only true if you know how compound interest works. In 2005 most of our students didn't and every single person on the Bigelow team got the compound interest question wrong.

4.1 Percent Applications

The percent part usually features short word problems. What's most important for these is to know how to go from the words to the equations. The most important word, of course, is percent. I assume you all know that this means divided by 100. For example, 15% means 0.15 and 175% means 1.75. The sections below discuss several other words you should know.

4.1.1 Of

The word "of" usually means times. For example, consider

Question: What is 15% of $40?

The answer is 0.15 × $40 = $6.

They also sometimes explicitly ask for you to give answers as reduced fraction instead of as a decimal. When they do this, the easiest way to get the answer is often to keep everything as a fraction and cancel common factors. For example,

Question: What is 175% of 16. Express you answer as a common fraction in lowest terms.

The easiest way to do this is to recognize that 175% = 1.75 = 7/4. Hence the answer is 16 × (7/4) = 4 × 7 = 28.

4.1.2 Larger than

The number that is "15% larger than x" is $x + 0.15x$ which is equal to $1.15x$.

Question: What number is 15% larger than 40?

For example the answer to the above question is 46. Problems involving numbers like 25% or 40% obviously work similarly. One thing they seem to like to do to try to confuse people is to percents like 175% in these problems.

Question: What number is 175% larger than 400.

The key here is to remember that the number 175% larger than x is $x + 1.75x$ which is equal to $2.75x$. The answer is $2.75 \times 400 = 1100$.

4.1.3 Percent off

Question: The original price of a pair of jeans at Delia's was $65. They are on sale at 40% off. How much would it cost to buy them?

Being sold for 40% off means they are sold for $65 – (0.4 × $65) = $39. You can also think of 40% off as meaning that they are sold at 60% of the regular price. Calculating 0.6 × $65 is another way to get the answer. IMLEM also likes to put sales taxes in problems. If there is a 5% sales tax then you have to add 5% to the total, i.e you need to multiply you answer by 1.05.

Question: The original price of a pair of jeans at Delia's was $65. They are on sale at 40% off. Caroline buys two pairs. How much does she need to pay in total if the sales tax rate is 5%?

The answer to this one is 1.05 × (2 × (0.6 × $65)) = $81.90. Problems in this category can usually be done pretty quickly, so you should be sure to read them carefully and go back and check your answers. It's easy to get questions wrong by noting that they asked "how much did she save" instead of "how much did she pay" or something like that. They also sometimes ask what the original price was. Here, you may want to use algebra to avoid getting confused.

Question: At its end-of-the-season sale Fast Splits is selling a pair of cross country skis for 20% off of the original price. If the sale price of the skis is $100 what is the original price?

The answer to this question is NOT $120. If the original price is x, then the sale price is $0.8\,x$. Hence, x is the solution to $0.8\,x = 100$, which implies $x = 100 / 0.8 = \$125$.

4.1.4 Percentage points higher

One confusing phrase that has come up occasionally is "percentage points higher." What this means is that you're supposed to subtract the two percentages in the problem. An example of this is:

Question: John goes to Lam's for dinner. The total bill for dinner is $50. John gives the waitress $60 and tells her to keep the change as a tip. How many percentage points higher than the standard tip of 15% was his tip?

Question (IMLEM Meet #4, March 2001): On each of the last two math tests Katherine got three questions wrong. One of those tests had 12 questions. The other had 37. How many percentage points higher was Katherine's score on the test with 37 questions than it was on the test with only 12 questions? Round your answer to the nearest percent.

John's tip is $10/$50 = 0.20 = 20%. The answer to the first question is 20 – 15 = 5 percentage points. Katherine's score on the 1st test was 9 out of 12 = 75%. Her score on the second test was 34 out of 37, which is approximately 91.89%. The answer is 91.89 – 75 ≈ 17 percentage points. Personally, I think these are really bad questions because the term is not very precise and too easy to get confused with similar wordings like "what percent larger than the standard tip of 15% was John's tip?" (The answer to this would be (20 – 15)/15 = 33% larger.)

4.2 Compound Interest

4.2.1 Basics

Question: Suppose you put $100 in the bank. At the end of each year the bank adds 10% interest to your account balance. How much do you have in the bank at the end of ten years?

Question: The average teacher salary for a public school teacher in Massachusetts is $56,000 per year. If the average salary increases by 5% per year, what will the average salary be ten years from now? Round your answer to the nearest dollar.

The answers to these questions are NOT $200 and $84,000.

The first problem involves what is called *compound interest*. The reason why the answer is not $100 is that you don't just get interest on the original $100. In every year after the first you also get interest on your interest. At the end of the first year you have $110. At the end of the second you have $110 + 0.1 × $110 = 1.1 × $110 = $121, i.e. you get $11 in interest in the second year because you're getting interest on $110 instead of $100. At the end of the third year you have $121 + 0.1 × $121 = 1.1 × $121 = $133.10. You could solve this problem on your calculator by just doing this over and over. It's quicker to recognize the pattern: each year you just multiply the previous total by 1.1. The answer is $100 × $(1.1)^{10}$ = $259.37.

- A general formula for compound interest problems is that if you start with $z and earn r % interest in each of T periods, then you end up with $z (1 + r/100)^{T}$.

For example, the answer to the second problem is $56,000 × $(1.05)^{10}$ = $91,218. An important reminder in these problems is that you should not round off until the very end. If you round off at the end of each year you get a slightly different answer, which would be marked wrong.

Here's one last practice problem:

Question (IMLEM Meet #4, Feb. 2002): When Shawn started his job in a furniture store in 1996 earning a salary of $31,500, he was promised a 4% annual raise. What was Shawn's salary in 2002? Round your result to the nearest dollar.

The answer to this one is $39,858. One mistake some people may have made on this one was to mistakenly multiply by 1.04^7 instead of 1.04^6. Although 2002 was his seventh year, he didn't get his first raise until 1997.

4.2.2 Annual interest rates

A potentially confusing aspect of many compound interest problems is that they talk about annual interest rates. For example,

> *Question (IMLEM Meet #4, Feb. 2006): Ivan deposited $2000 in a bank account that pays 4.8% annually and compounds monthly. How much money can Ivan expect to earn in interest if the certificate of deposit is for 12 months? Round your answer to the nearest whole number of dollars.*

> *Question (IMLEM Meet #4, Feb. 2003): Stephanie deposited $10,000 at a bank in an 18-month certificate of deposit account that pays an annual interest rate of 3%. How much money can Stephanie expect to have in her bank account at the end of the 18 months if the interest is compounded monthly? Express your answer to the nearest whole number of dollars.*

The 4.8% "annual interest" mentioned in the first question means that the monthly interest rate is one-twelfth of 4.8% or 0.4%. The monthly interest rate is what you need to plug into the formula I gave above. At the end of 12 months Ivan has $2000 \times (1.004)^{12} = \$2098$. The amount he earned in interest is $98. (This was an example of a question where you had to read carefully and realize that they asked for the interest earned instead of the final balance.)

The 3% "annual interest rate" in the second problem corresponds to a monthly interest rate of one-twelfth of 3% or 0.25%. The answer is $10,460.

4.2.3 Advanced topic: continuous compounding

In the problem above, if you carry the results out to the nearest cent you find that Ivan gets $98.14 in interest instead of $96.00 because the interest is compounded monthly. What would he have gotten if the interest were instead compounded daily?

The daily interest rate would be $4.8\%/365 \approx 0.013150685\%$. Hence, the amount he'd have after 365 days is $\$2000 \times 1.00013150685^{365} = \$2098.33$, which is a little more.

Could Ivan do any better if he convinced the bank to compound the interest more often? Sure he could. The bank could compound his interest hourly instead of daily. I won't bother to write out the miniscule hourly interest rate, but if you do you'd find that he'd make $98.34 instead of $98.33.

Banks have noticed that once you're compounding interest often enough it really doesn't cost much more to compound it even more often. Hence, if you want to sound really generous (without actually spending much) you can offer to compound daily, hourly, or even every second.

Is there a limit to this process? It turns out there is and you can find it on your calculator. If you divide the year into n periods, Ivan's end of year balance would be $2000 \times (1 + 0.048/n)^n$. There's a theorem that says that if you evaluate this expression as n gets larger and larger you find that the answer gets closer and closer to $2000\, e^{0.048}$, where e is Euler's constant.

If you have a good calculator it probably has a button that will let you calculate this. It might be labeled e^x. It might be labeled *exp*. To check whether you've found the right button try entering the number one and then pressing the button. If it's the right one it should give you the value of e, which is approximately 2.71828182845904523536. Like π, e is a transcendental number. It's not only irrational, but there's no polynomial equation that has e as a root. There are lots of infinite formulas for it that let you approximate it accurately. The best known of these is

$e = 1/0! + 1/1! + 1/2! + 1/3! + 1/4! + 1/5! + \ldots$

(The ! symbols in this equation are factorial symbols.)

Going back to the problem above you can check on your calculator that $2000\, e^{0.048}$ is also 2098.34 (when rounded to the nearest cent). This means that if Ivan's bank compounds the interest every second or even every millisecond, he'll only do at most a cent better than if they compound the interest daily.

Giving compound interest according to the formula with an e in it is called *compounding continuously*. Lots of banks do it.

Category 5 – Algebra

The algebra category in meet #4 covers word problems. Officially it's "word problems (linear, including direct proportions and systems)." As always, I can't really teach algebra, but will try to summarize the kinds of problems that are included and give a few tips.

5.1 Functions

Word problems about functions often describe a function *f(x)* verbally and then ask you to evaluate it for a given value of *x*. For example,

> *Question: Elizabeth's cell phone plan gives her up to 300 minutes a month for a fixed charge of $29.95. She pays an extra 35 cents per minute for all minutes beyond 300. How much will she pay in a month if she makes 430 minutes of calls?*

The discussion of the cell phone plan is verbally describing the function:

$$f(x) = \begin{cases} 29.95 & \text{if } x < 300 \\ 29.95 + 0.35(x - 300) & \text{if } x \geq 300 \end{cases}$$

The answer is $f(430) = 29.95 + 0.35(130) = \$75.45$. Many real-world cell phone plans work this way. You can end up paying a lot more if you go over your minute allotment by something that doesn't seem so big.

Some problems also ask you to go the other way: they give you the value of the *f(x)* and ask you to find *x*.

> *Question: Elizabeth's cell phone plan gives her up to 300 minutes a month for a fixed charge of $29.95. She pays an extra 35 cents per minute for all minutes beyond 300. One month she wasn't careful to keep track of her minutes and got a bill for $63.90. How many minutes did she use?*

In this problem you know that she used more than 300 minutes. Hence, the answer x is the solution to $29.95 + 0.35(x - 300) = 63.90$. The answer is 397.

5.2 Two Equations in Two Unknowns

There are many ways to make up word problems that require you to solve two linear equations in two unknowns. For example,

Question: The Fishers raise both sheep and chickens on their farm. In total their farm animals have 73 heads and 270 legs. How many chickens do they have?

Question: The Underwood bake sale sells cookies and brownies. George brings $3.30 to spend at the bake sale. He notices that there are two ways he could spend all of his money: he could buy 2 brownies and 10 cookies; or he could buy 7 brownies and 2 cookies. How much does a cookie cost?

Question: Three years ago, Anna was half as old as her sister Caroline is now. If Caroline is three years older than Anna now, how old will Anna be in four years?

Writing s for the number of sheep and c for the number of chickens, the first question is describing the system:

$$s + c = 73$$
$$4s + 2c = 270$$

One way to solve this is to multiply the first equation by 4 to get $4s + 4c = 292$, and then to subtract the second equation from the first to get $2c = 22$. The answer is 11.

The second question is describing the system:

$$2b + 10c = 330$$
$$7b + 2c = 330$$

The answer is that a cookie costs 25 cents. (A brownie costs 40 cents.)

The third question is describing the system

$$(A - 3) = \tfrac{1}{2} C$$
$$C = A + 3.$$

One thing to be careful of here is to notice that they're asking you to find $A + 4$ rather than A. The answer is 13. This one is easy enough that you might be tempted to do it without algebra. This is fine if you don't get confused. I usually think it's safer to write down the equations. In this case I'd solve by substituting the second equation into the first to get $(A - 3) = \tfrac{1}{2} (A + 3)$ and then working from there.

5.2.1 Problem solving tip: Know your category

A problem solving tip is that the category is linear equations so if you get something complicated that looks nonlinear there is probably some trick that makes the question easier. For example if you're given the system:

Question: Suppose $\dfrac{A-1}{B-1} = \dfrac{3}{2}$ *and* $\dfrac{B+1}{2A} = \dfrac{3}{5}$. *Find A*

The first thing you should think in a question like this is "what is this doing in a category on linear equations." Hopefully, if you do this the next thing you'll think is "Oh. If I multiply both sides of the left equation by 2(B-1) and both sides of the right equation by 10A these will be standard linear equations." People sometimes call this cross-multiplying. The first equation becomes 2(A-1) = 3(B-1). The second becomes 5(B+1) = 6A. The answer turns out to be A=2.5.

145

Here's another example from an IMLEM Meet:

Question (IMLEM Meet #4, Feb. 2005): In parallelogram MATH, MA=168mm, AT=15x + 3mm, TH=3xy^3 mm, and MH = 108mm. What is the numerical value of y^x?

The one fact you need to know about parallelograms is that opposite sides are equal. Hence, the word problem is just giving you the system:
$$3xy^3 = 168$$
$$15x+3 = 108.$$
Again, you should think that the topic is not solving cubic equations so this can't really be necessary. In this case, notice that you can immediately solve the second equation and find $x=7$. Plugging this into the first equation gives $21y^3 = 168$ or $y^3=8$ so $y=2$. The answer is $2^7 = 128$.

5.3 N equations in N unknowns

Occasionally, they also ask problems with more unknowns. For example, the following problem has 3 unknowns:

Question (IMLEM Meet #4, Feb. 2005): Together Jim and Bob weigh 357 pounds. Together Jim and Larry weigh 393 pounds. The combined weight of all three men is 565 pounnds. How much do Bob and Larry weigh together.

This problem asks you to solve the system:
$$J + B \quad\;\; = 357$$
$$J \quad\;\; + L = 393$$
$$J + B + L = 565$$
In general, the way to solve 3 equations in 3 unknowns is always to reduce them to a 2-equations-in-two-unknowns problem. (And the way to solve 4 equations in 4 unknowns is to reduce it to a 3-equations-in-3-unknowns problem, and so on). There are two ways to do this reduction:

Method 1 is to find one of the unknowns you can solve for and then substitute in for this unknown in two of the equations. For example, the first equations gives $B = 357 - J$, so we can plug this into the second and third equations to get the two equation system:
$$J + L = 393$$
$$J + (357 - J) + L = 565$$
This turns out to be a pretty easy system to solve.

Method 2 is to eliminate one of the unknowns from the equations in two different ways: once by subtracting some multiple of equation 2 from equation 1; and once by subtracting some multiple of equation 3 form equation 1. In this case, the question asks for B + L, so the natural thing is to eliminate J. Subtracting the second equation from the first gives
$$B - L = -36$$

Subtracting the third equation from the first gives
$$- L = -208.$$
Again, this two equation system is easier to solve than we had any right to expect, but why complain?

However you do it, the answer should be 380.

5.3.1 Problem solving tip: You may not need to solve the equations

The fact that each of the approaches I took to the above problem gave a very easy set of equations is probably not a coincidence. IMLEM must recognize that it would be unreasonable to ask you to solve a complicated three equation system in just three minutes. Because of this, if you see what looks like a complicated set of equations to solve the trick may be that you don't need to solve them all. Here are a couple examples:

Question: Larry, Kevin, Robert and Danny divide up a dozen cookies. Larry gets as many as Kevin and Robert combined. The difference between the number that Larry gets and the number that Kevin gets is two more than the difference between the number that Robert gets and the number that Danny gets. Together, Robert and Danny get one more than Kevin. What is the average of the number of cookies that Larry gets, the number of cookies that Kevin gets, the number that Robert gets, and the number that Danny gets?

Question (IMLEM Meet #4, Feb. 2006) At the Useless Trinkets Gift Shop, 4 blips and 5 clips cost 87 cents, 2 clips and 6 glips cost 52 cents, and 3 blips and 1 glip cost 29 cents. How many cents would 2 blips, 2 clips and 2 glips cost?

The first problem is designed to make you start writing down equations frantically: $L + K + R + D = 12$, $L - K = R - D + 2$, and so on. The trick is that this is completely unnecessary. The question doesn't ask you to solve for how many cookies each person got. It only asks you to find the average. Everything you need to know is in the first sentence! The average is 3 regardless of how they divide up the twelve cookies.

The second question is just a little harder. It describes the system of equations:
$$
\begin{aligned}
4b + 5c \quad &= 87 \\
2c + 6g &= 52 \\
3b \quad + \ g &= 29
\end{aligned}
$$
The fact that this is a fairly complicated 3 equation system and that they ask for a particular combination of the three prices are tipoffs that there may be a trick. In this case, the trick is to add the three equations together. This gives $7b + 7c + 7g = 168$. Dividing by 7 and then multiplying by 2 gives that the answer is 48.

You could also have used this tip on the problem about Jim, Bob, and Larry I discussed in the previous section. Not from the very start, but once you got to the point where you knew that $B - L$ was -36 and $-L$ was -208, you could have finished it off quicker by not solving for B. Instead, you just write $B + L = B - L + 2L = (B - L) - 2(-L) = -36 + 416 = 380$.

5.4 Time-Distance-Speed Problems

Problems about average speeds are popular in both math classes and math meets because people who try not to use math to solve them often get them wrong. A typical example is:

Question: Jessica lives three miles from school so she usually rides her bike. One morning she starts out riding at a speed of 9 miles per hour. When she gets half way to school, she gets a flat tire and has to walk the rest of the way. If she walks at 3 miles per hour, what is her average speed for the trip.

The answer to this question is NOT 6 miles per hour. The reason is that Jessica spends much more time going 3 miles per hour. Traveling 1½ miles at 9 mph takes one-sixth of an hour or ten minutes. Traveling 1½ miles at 3 mph takes 30 minutes. If I'd phrased the question as "Jessica travels for ten minutes and 9 mph and then for 30 minutes at 3 mph" you would not have been tempted to say 6 mph.

The correct answer is pretty simple. All together she goes 3 miles in 40 minutes (which is 2/3 of an hour). Her average speed is 3 / (2/3) = 4½ miles per hour.

The three main things to remember in doing problems like these are:

- Speed = Distance / Time
- Time = Distance / Speed
- Distance = Speed × Time

Something that makes them relatively easy to remember is that they're just three versions of the same formula.

My main piece of advice for doing these problems is to focus on finding the distance or the time, not on the speed.

Question (IMLEM Meet #4, Feb. 2006): Eliot drives to work by a 12 mile route and averages 30 miles per hour. He drives home by a 15 mile route and averages 40 miles per hour. What is his average speed for the entire round trip of 27 miles? Give your answer to the nearest tenth of a mile per hour.

Question (IMLEM Meet #4, March 2001): Tom must travel 90 miles and hopes to average 60 miles per hour. Unfortunately he gets stuck in traffic early on and averages only 30 mph for the first 36 miles. What speed must Tom average for the remainder of the trip so that he still averages 60 mph for the entire trip. Express your answer in miles per hour to the nearest whole number.

In the first problem they give you the distances, so the thing to do is to find the times. Driving 12 miles at 30 mph takes 12/30 = 2/5 of an hour. Driving 15 miles at 40 mph takes 15/40 = 3/8 of an hour. The total driving time is 2/5 + 3/8 = 31/40 of an hour. The average speed is 27 / (31/40) ≈ 34.8 miles per hour.

The second one is similar. Traveling 36 miles at 30 mph takes 36/30 = 6/5 hours. To average 60 miles per hour for the whole 90 miles Tom must finish the trip in 90/60 = 3/2 hours. Hence, he must do the remainder of the trip in 3/2 – 6/5 = 3/10 of an hour. This will be very hard for Tom. To go the remaining 54 miles in just 3/10 of an hour his speed will need to be 54 / (3/10) = 180 miles per hour! This is impossible in most cars. Don't try it at home: it's unsafe and illegal throughout the U.S.

One special case that comes up regularly is problems where someone travels the same distance at to different speeds. My first problem was one example. Other examples are problems where someone goes to work/school/a friend's house at one speed and then returns via the same route at a different speed. Doing some simple algebra shows:

- If you travel x miles at speed s_1 and then another x miles at speed s_2, then your average speed for the full trip is $\dfrac{2s_1s_2}{s_1 + s_2}$.

This expression is called the harmonic mean of s_1 and s_2.

Knowing that the average speed in a constant-distance problem is the harmonic mean is not only useful for answering questions where they ask you for the average speed. Consider

Question (IMLEM Meet #4, Feb. 2002): Julie left the house at noon and rode her bicycle at an average speed of 15 miles per hour until she got a flat tire. She then walked home by the same route at an average speed of 3 miles per hour. If she arrived home at 4:00pm, how far away was she when she got the flat tire?

Using the harmonic mean formula you could know immediately that her average speed had been 2 ×15 × 3 / (15 + 3) = 5 miles per hour. Traveling for four hours at an average speed of 5mph you go 20 miles. Hence, she was half this far away, 10 miles, when she got the flat tire.

Of course, you could have also done this by the standard method using a little algebra. If she was x miles away when she got the flat then she would have spent x/15 hours on the way out and x/3 on the way back. The solution to x/15 + x/3 = 4 is x = 10.

5.4.1 Advanced topic: arithmetic, geometric, and harmonic means

The harmonic mean is one of three notions of an average that come up frequently. These are:

- The arithmetic mean of *a* and *b* is $(a + b)/2$.

- The geometric mean of *a* and *b* is $\sqrt{ab}$.

- The harmonic mean of a and b is $\dfrac{2ab}{a+b}$.

The harmonic mean is the appropriate concept to use when talking about average speeds in a constant distance problem. The geometric mean is appropriate for other problems (one of which is computing average interest rates in problems with compounding.) Arithmetic means, of course, come up all the time.

You may have noticed that in the constant-distance problems I discussed the average speed was always less than the arithmetic mean of the two speeds. This is not a coincidence. A well known theorem relating the three types of means is:

Theorem: If a and b are positive numbers then

$$\min(a,b) \le \frac{2ab}{a+b} \le \sqrt{ab} \le \frac{a+b}{2} \le \max(a,b).$$

In words this says that the arithmetic mean is always at least as big as the geometric mean, which is at least as big as the harmonic mean.

The proof is surprisingly easy. The arithmetic-geometric mean comparison is:

$$\frac{a+b}{2} \ge \sqrt{ab} \quad \Leftrightarrow \quad (a+b)^2 \ge 4ab$$

$$\Leftrightarrow \quad a^2 + 2ab + b^2 \ge 4ab$$

$$\Leftrightarrow \quad a^2 - 2ab + b^2 \ge 0$$

$$\Leftrightarrow \quad (a-b)^2 \ge 0$$

The geometric-harmonic comparison is an immediate corollary. The product of the arithmetic mean and the harmonic mean is the square of the geometric mean. Hence, if the arithmetic mean is greater than the geometric, it must be that the harmonic is less than the geometric.

$$\frac{a+b}{2}\frac{2ab}{a+b} = \sqrt{ab}\sqrt{ab} \quad and \quad \frac{a+b}{2} \ge \sqrt{ab} \quad \Rightarrow \quad \frac{2ab}{a+b} \le \sqrt{ab}$$

IMLEM Meet #5

Meet #5 is a fitting close to the IMLEM season. They really step up the amount of material covered in each category. More of the topics are things you don't see much in middle school. The meet #5 section is longer than any of the previous ones. As you read it, you'll also notice that it doesn't have any comments saying things like "most of what you need to know for this category is what's on page 5." The arithmetic category alone could be the subject of a full semester class. The algebra category seems to be the easiest one, but only if you know algebra.

You are allowed to use a calculator in meet #5. For the arithmetic category, it will be an advantage to have one that does combinations and permutations.

Category 2 – Geometry

The geometry problems in meet #5 are about spatial geometry. This is a not a great topic for a middle school math meet: you need to memorize a whole bunch of formulas and for a lot of them there isn't an easy way to explain why they are true. Still, it's the topic IMLEM chose, so here it goes.

2.1 Surface Areas and Volumes

Most meet #5 questions are about surface areas and volumes.

2.1.1 Volumes

- The *volume* of a three-dimensional shape is the amount of cubic units of space contained in it.

One can make up volume problems by just asking directly about volumes, e.g.

Question (IMLEM Meet #5, Apr. 2004): A roll of toilet paper has a diameter of 12cm and a height of 11.4cm. The inner diameter of the tube is 4cm. How many cubic centimeters are in the volume of the solid part of the roll of toilet paper?

There are also a variety of types of word problems that essentially ask you to compute volumes. One is *displacement* problems. If you drop a sphere, cube or other three-dimensional object into a completely full cup or water (or any other liquid), the volume of liquid that spills out is equal to the volume of the object you dropped in. (This is something that people actually do to find the volume of irregularly shaped objects.) For example,

Question (IMLEM Meet #5, Apr. 2004): A steel sphere is dripped into a full cubic decimeter, displacing some water, which spills over the sides. When the ball is removed it is noted that 732 milliliters of water remain in the cube. How many centimeters are in the diameter of the sphere?

The first step in doing this problem is to note that the volume of the sphere must equal the volume of water displaced: 1000cc – 732ml = 268cc. (One cubic centimeter is the same as one milliliter). The second step is to use the formula for the volume of a sphere to figure out what the radius must be for the sphere to have this volume.

Another type of problem where you'll want to use volumes is questions about how many blocks would be needed to build something. For example,

Question (IMLEM Meet #5, Apr. 2001): How many 8-inch by 8-inch by 16-inch cement blocks will be required to construct a foundation that is 32 feet long, 24 feet wide and 8 feet tall. Each wall is 8 inches thick.

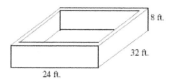

The way you use volumes here is to first figure out the volume of the foundation, then figure out the volume of an 8 × 8 ×16 block and then divide the two to get the number of blocks needed. One thing to watch out for in this problem is that some dimensions are given in inches and others in feet. Another tricky aspect is that you have to be sure to account for the thickness of the wall so you don't double-count the corners, i.e. the answer is not the same as it would be for a straight wall that was 112 feet long.

2.1.2 Surface area

- The *surface area* of a solid is the total number of square units of space on boundary of the solid. For example, the surface area of a cube is the sum of the areas of the six squares that are its sides.

The most common problems for which you need to know surface area are how-many-cans-of-paint-to-buy problems, e.g.

Question: The moon is approximately a sphere with radius 1080 miles. If one can of spray paint covers ten square feet, how many cans of spray paint would be needed to paint the surface of the moon?

To do problems like these, of course, you don't just need to know what "volume" and "surface area" mean. You also need to know formulas for the volume and surface area of the shapes they ask about. They don't tell you the formulas on the meets, so you'll need to do a fair amount of memorizing.

2.2 Shapes Related to Circles

2.2.1 Spheres

A *sphere* is a completely round shape that looks like a ball. Formally, a sphere with radius r is the set of points at a distance of r from one point, which is called the center of the sphere. Mathematicians usually use the word sphere to refer only to the surface of the three dimensional shape and use the word *ball* to refer to the whole solid shape including the interior, but sometimes people use the word sphere for that too.

The two main facts to memorize are:

- The volume of a sphere of radius r is $V=(4/3)\,\pi\, r^3$.

- The surface area of a sphere of radius r is $A=4\,\pi\, r^2$.

2.2.2 Hemispheres

A *hemisphere* is half of a sphere.

You could memorize the volume and surface area facts, but you should be able to figure them out right away if you remember the sphere facts.

- The volume of a hemisphere of radius r is $V=(2/3)\,\pi\, r^3$.

- The surface area of a hemisphere of radius r is $A=2\,\pi\, r^2 + \pi\, r^2 = 3\,\pi\, r^2$

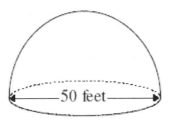

Be careful about what exactly is being asked in questions about surface area. The reason why the surface area of a hemisphere is $3\,\pi\, r^2$ is that the round part on the top has half the surface area of a sphere, $\frac{1}{2}(4\,\pi\, r^2) = 2\,\pi\, r^2$, and the bottom is a circle, which has area $\pi\, r^2$. For example, in 2001 they asked

Question: (IMLEM Meet #5, Apr. 2001): The exterior surface of a hemispherical capital dome is to be covered with gold leaf. If the diameter of the dome is 50 feet, how many square feet of gold leaf will be required to cover the dome.

It is implicit in the verbal description that the gold leaf is only to be placed on the upper round part of the hemisphere, not on the flat bottom part (which wouldn't be there in the typical dome). Hence, the problem is asking you to compute $2\pi r^2$ for $r = 25$.

2.2.3 Cylinders

A *cylinder* is shape that looks like a can. The top and bottom faces are circles of radius r. The center of the top face is directly above the center of the bottom face. The height h of the cylinder is the distance between the top face and the bottom face.

- The volume of a cylinder is $V = \pi r^2 h$.

- The surface area of a cylinder $A = 2\pi r h + 2\pi r^2$

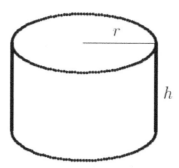

The area and volume formulas for a cylinder are easier to understand that the sphere formulas. To think about the volume formula, just think of filling up the cylinder with a stack of coins each of which has radius r and height 1. To think about the surface area formula, note that you'll make a cylinder of radius r and height h if you take a standard rectangular piece of paper with width $2\pi r$ and height h, and tape the left and right sides together.

A sample cylinder problem is:

Question (IMLEM Meet #5, Apr. 2001): The label from each of the two cans shown here is a 4-inch by 12-inch rectangle which exactly covers the rounded surface of each can. If the height of can A is 4 inches and the height of can B is 12 inches, what is the ratio of the volume of can A to the volume of can B?

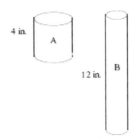

The answer to this question is 3:1. You can see this without taking the time to compute the volume of either cylinder by noting that the height h_A of cylinder A is one-third of the height of cylinder B, and the radius r_A of cylinder A is three times as large as the radius of cylinder B. Hence, $V_A = \pi r_A^2 h_A = \pi (3 r_B)^2 (\frac{1}{3} h_B) = 3V_B$.

If you did want to compute the volume of cylinder A you'd use a two-step approach: first use the fact that $2\pi r_A = 12$ to find r_A: and then plug this into the volume formula.

Again, be careful about what is being asked in questions about surface area. The $2\pi r h$ part of the surface area is the area of the round part of the cylinder that would be covered by a label on a food can. The $2\pi r^2$ part of the surface area is the area of the top and bottom faces.

If you look at a cylinder from above it looks just like a circle of radius r. If you look at it from the side it looks just like a rectangle with width $2r$ and height h.

2.2.4 Cones

A cone is a shape that looks like an ice cream cone. The base is a circle of radius r. It narrows to a point (called the vertex), which is located at a height of h directly above the center of the base.

- The volume of a cone is $V = (1/3)\, \pi r^2 h$. Note that this is one-third of the volume of a cylinder of the same height. You can think of this as being like the (1/2) *base × height* formula for the area of a triangle, but with a one-third in it instead of a one-half because it's a three dimensional figure instead of a two dimensional figure.

- The surface area of a cone is $A = \pi r \sqrt{r^2 + h^2} + \pi r^2$. Again, the first part of this formula is the area of the curved surface. The second part is the area of the flat circle on the bottom of the cone. You also see the formula written as $A = \pi r s + \pi r^2$, where s is the *slant height*, the distance along the surface of the cone from the vertex to a point on the outside of the bottom circle.

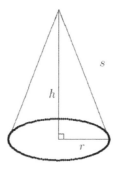

From the top a cone looks like a circle. From the side it looks like an isosceles triangle with base $2r$, two sides of length s, and height h.

There is a neat explanation for why surface area of the top part of a cone is $\pi r s$. It involves thinking of how you can get a cone by folding up a sector of a circle of radius s. (You get a cone this way because the vertex of a cone is equidistant from each of the points on the bottom circle.) If you have some time you could try to figure out how to make a cone using a pencil, ruler, compass, scissors, and tape.

A typical cone problem is

> *Question (IMLEM Meet #5, Apr. 2005): A solid cone with radius 2.25 cm and height 4.5 cm is placed inside a cylinder that also has radius 2.25 cm and height 4.5 cm. How many cubic centimeters of space inside the cylinder are not occupied by the cone?*

The answer to this is $\pi \, 2.25^2 \, 4.5 - \frac{1}{3} (\pi \, 2.25^2 \, 4.5) = \frac{2}{3} (\pi \, 2.25^2 \, 4.5)$. This problem is not exactly the same as the original. An unfortunate aspect of many meet #5 geometry problems is that they ask you to substitute 3.14 for pi and round your answer to the nearest tenth. I can't quite see what the point of this is, but it comes up a lot so be sure to practice typing quickly on your calculator.

Archimedes knew the volume formulas for cones and spheres in 200 BC, but it's hard to give a rigorous explanation without using calculus. There is a very nice way to see why the formula for a sphere works once you know the formulas for cylinders and cones. I describe it in the Advanced Topics section.

2.3 Prisms and Pyramids

2.3.1 Prisms

A prism is kind of like a cylinder, but with polygons as the top and bottom faces instead of circles. As with a cylinder, we write h for the distance between the top and bottom face.

A cube is one example of a prism. So is the pentagonal prism shown below. (It's shown lying down. To see that it's a prism think of it as standing up on one of its pentagonal ends.)

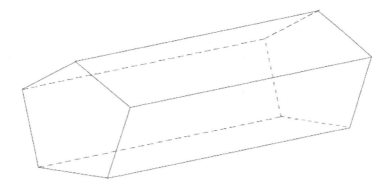

- The volume of a rectangular prism is $V = A_{base} h$, where A_{base} is the area of the base.

- To figure out the surface area of a prism you figure out and add up the areas of the top, bottom, and all faces on the sides. The sides are always rectangles. The top and bottom can be any polygon.

- If the top and bottom of the prism are regular *n*-gons with sides of length *s*, then the surface area will be $nsh + 2 A_{base}$. Figuring out A_{base} is hard unless the n-gon is triangle square, or hexagon. For these A_{base} is $s^2 \sqrt{3} / 4$, s^2, and $s^2 3\sqrt{3} / 2$.

2.3.2 Pyramids

A pyramid is solid that has a polygon as its base and shrinks to a single point at a point some distance above the center of the base. As with a cone, we write *h* for the distance between the bottom face and the top vertex.

The classic example is the kind of square pyramid like the one the Ancient Egyptians liked to build. Another example is a tetrahedron. You can make a pyramid by folding up a piece of paper like the one shown below.

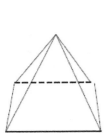 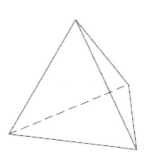 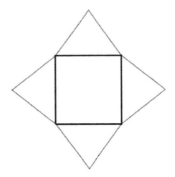

- The formula for the volume of a pyramid is just like the formula for the volume of a cone: $V = (1/3) A_{base} h$.

If you aren't told the height of the pyramid you can sometimes figure it out using the Pythagorean Theorem. For example, in the piece of paper on the right above that's going to be folded up to form a pyramid, suppose the square and the triangles all have sides of length s. First, you can use the Pythagorean Theorem (or your knowledge of equilateral triangles) to say that the slant height of each of the sides will be $s\sqrt{3}/2$. Then you use the Pythagorean theorem again to say that the height of the pyramid satisfies: $h^2 + (s/2)^2 = (s\sqrt{3}/2)^2$. This gives $h = s/\sqrt{2}$.

Rather than deriving this during a math meet you could just remember that the volume of a regular rectangular pyramid is $V = s^3/(3\sqrt{2})$.

While you're at it you could memorize that the volume of a regular tetrahedron with side s is $V = s^3\sqrt{2}/12$.

These formulas only work for regular pyramids which have all edges of the same length. For most problems on the math meets I'd guess that it will be enough to just remember the basic $V = (1/3) A_{base} h$ formula.

- To figure out the surface area of a pyramid you figure out and add up the area of the bottom and the areas of all the faces on the sides. The sides are always triangles. The bottom can be any polygon.

In a regular tetrahedron the surface area is $A = s^2\sqrt{3}$.

In a square pyramid with a square of side s as the base and height h the surface area is $A = s^2 + s\sqrt{s^2 + 4h^2}$. For a square pyramid with all edges of length s this simplifies to $A = s^2(1 + \sqrt{3})$.

2.4 Polyhedra

A polyhedron is a three dimensional shape that has polygons for all of its faces. We've already seen two types of polyhedra: prisms and pyramids.

2.4.1 Regular polyhedra

There are five regular polyhedra – polyhedra in which all faces are identical regular n-gons and each vertex has an identical configuration of faces around it. They are:

Tetrahedron – four triangular sides
Cube – six square sides
Octahedron – eight triangular sides
Dodecahedron – twelve pentagonal sides
Icosahedron – twenty triangular sides.

The figures below show what they look like, and how you could get each one by folding up a piece of paper. (Note that the fold-ups are not below the corresponding polyhedron. Can you figure out which makes which shape?)

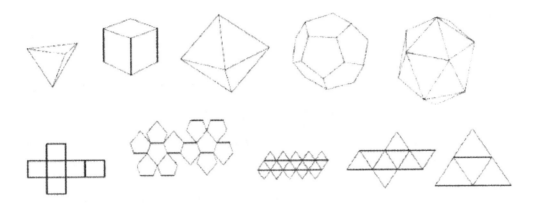

2.4.2 Vertices, edges, faces

- The points at the corners of a polyhedron are called *vertices*. For example, a cube has eight vertices: four on the bottom and four on the top.

- The polygons that make up the outside of a polyhedron are called *faces*. For example, a cube has six square faces.

- The line segments that are the borders of the faces in a polyhedron are called *edges*. For example, a cube has 12 edges.

You could be asked to count the number of edges in some figure, e.g.

Question: How many edges does a cube have?

It's not a very exciting question, but one thing that's kind of nice about questions like this is that there are three different methods one can use to solve them:

- o Picture the shape. For example, a cube has four edges in the bottom square, four in the top square, and four vertical edges connecting the bottom square and the top square.

159

o Look at vertices. A cube has 8 vertices. Each vertex has three edges coming out of it. 8 × 3=24. Each edge has two endpoints, so you divide this by 2 and get 12.

o Look at faces. A cube has 6 faces. Each has four edges. 6 × 4 = 24. Each edge belongs to two faces, so again you divide this by 2 and get 12.

The one other result you need to know relating vertices, edges, and faces is <u>Euler's Theorem</u>:

• Suppose a polyhedron has no holes. Let V be the number of vertices, E the number of edges and F the number of faces. Then, $V - E + F = 2$.

Euler's Theorem can be useful in a couple different situations. One of these is in doing problems where you're having trouble picturing the figure. For example,

Question: An icosahedron has twenty triangular sides. How many vertices does it have?

Even if you don't know what an icosahedron looks like you should be able to figure out that F = 20 and E = 30. They tell you F in the question, and you can count the edges by saying that each of the 20 sides has 3 edges, but that 20 × 3 is double-counting each edge. Euler's Theorem then gives V − 30 + 20 = 2, which implies V = 12.

In 2001 the Mystery category included two spatial geometry problems. One was

Question: Let V equal the total of the number of vertices on the tetrahedron, the hexahedron and the octahedron shown below. Let F equal the total of the number of faces and let E equal the total of the number of edges on the three solids. Find V + F − E.

If you didn't know Euler's Theorem you'd need to count up the vertices, edges, and faces of each solid. If you do, it's immediately obvious that the answer is 6: $V_i - E_i + F_i = 2$ for each of the three solids. Why they decided to call the cube a hexahedron I'll never know.

2.4.3 Diagonals

IMLEM seems to like to ask questions that involve counting diagonals in polyhedra. Perhaps this is because the reasoning needed to do them is similar to what's covered in the arithmetic category.

Line segment connecting two vertices of a polyhedron can be classified into three groups:

• Edges

- Surface diagonals. These are segments that are not edges of the polyhedron, but connect two vertices that belong to the same face.

- Space diagonals. These are segments that connect two vertices that do not belong to the same face. Unlike the others, they pass through the interior of the polyhedron.

One result relating the number of each of the three is

- $E + SurfaceDiag + SpaceDiag = V(V-1)/2$.

This is easy to see. One can draw a segment by picking any one of the V vertices and connecting it to any of the other $V-1$ vertices. This is counting each segment twice, however, so the total number of segments one can draw connecting vertices is $V(V-1)/2$.

Counting surface diagonals is not usually very hard. You can figure out the number of surface diagonals of a polyhedron by figuring out how many faces of each shape the polyhedron has. For example, a pentagonal prism has 2 pentagons and 5 rectangles as faces. In planar geometry an n-gon has $n(n-3)/2$ diagonals. A pentagon has 5 diagonals. A rectangle has 2 diagonals. $2 \times 5 + 5 \times 2 = 20$. The general formula for an n-gon prism is $n(n-3) + 2n = n(n-1)$. In an n-gon pyramid the answer is $n(n-3)/2$.

Counting space diagonals can be more confusing. If a figure is regular the formula is $VW/2$, where W is the total number of vertices that have no face in common with a given vertex. (For example in a cube $V = 8$ and $W = 1$ because the only vertex that doesn't share a face with a given vertex is the vertex at the opposite corner.) If you can't figure it out from thinking what the figure looks like you can always count the number of edges and surface diagonals and then use the formula above to find the number of space diagonals by subtraction.

Question (IMLEM Meet #5, Apr. 2004): How many space diagonals are there in a hexagonal prism such as the one shown below?

The direct way to do this is to say that you can make a space diagonal by connecting any of the six points on the left side with the three points on the right side that do not share an edge or a face: $6 \times 3 = 18$.

The indirect way to do it would take longer, but gives you a way to check your work. The figure has 18 edges: six on the left hexagon; six on the right hexagon; and six connecting the two. There are 30 surface diagonals: nine on each hexagonal face and two on each rectangular face. The number of surface diagonals can thus be computed as $V(V-1)/2 - 18 - 30 = 66 - 48 = 18$.

In an *n*-gon prism the number of space diagonals is *n (n-3)*. In an *n*-gon pyramid it's zero.

2.5 Units of Measurement

A common (and frustrating!) source of mistakes in surface area and volume questions is getting the units wrong. Sometimes you just happen to make mistakes. Other times the questions are designed to make you make them, e.g. by giving some distances in inches and others in feet.

The easiest units of measurements questions just ask you to convert distances, e.g

Question: The scale on a map is ½ inch = 1 mile. What is the distance between two cities that are 2.25 inches apart on the map?

One principal to keep in mind when converting between units is that distances scale linearly, surface areas scale quadratically, and volumes scale cubically. For example, in the metric system we have:

one meter = 100 centimeters
one square meter = 100^2 square centimeters = 10,000 square centimeters
one cubic meter = 100^3 cubic centimeters = 1,000,000 cubic centimeters

The same principal applies when doing English to metric conversions, or when converting between different English units, e.g.

one inch ≈2.54 centimeters
one square inch ≈$(2.54)^2$ square centimeters ≈ 6.45 square centimeters
one cubic inch ≈$(2.54)^3$ cubic centimeters ≈ 13.387 square centimeters

one square yard = 3^2 square feet = 9 square feet.

A fact about the metric system that's good to know is that a milliliter is the same as one cubic centimeter. Note that this implies that a liter is a cubic deciliter. A cubic meter has a volume of one thousand liters!

Metric units for weights and volumes are also related: one milliliter of water weighs one gram.

2.6 Advanced Topics

There's enough material on meet #5 so that I imagine few of you will want to read this extra section, but I figured I'd mention a few things just in case.

2.6.1 Where does the sphere formula come from?

The way the argument goes is to compare a hemisphere of radius r with a cylinder of radius r and height r that has had a cone cut out of it.

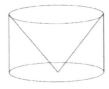 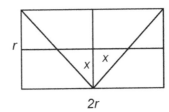 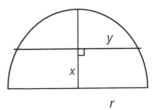

Looking from the side, the cylinder minus cone will look like a rectangle with a triangle cut out of it and the hemisphere will look like a semicircle. The neat way to find the volume of the hemisphere is to argue that the two shapes have the same volume. The way to see this is to think about the area of the cross-section at each height x above the base. In the cylinder minus cone the cross sections are disks with circular holes cut out. The area of the disk is always πr^2. The area of the hole is πx^2. So the total area of the cross section is $\pi (r^2 - x^2)$. In the case of the hemisphere the Pythagorean theorem gives $y = \sqrt{r^2 - x^2}$, so the area of the cross-section is πy^2 $= \pi (r^2 - x^2)$. Because the two shapes have the same cross-section area at each height, they have the same volume.

Computing the volume of the sphere is now easy. The volume of the cylinder is $\pi r^2 r$. The volume of the cone is minus cone is $1/3 \, \pi r^2 r$. So the volume of the hemisphere is $\pi r^3 - 1/3 \, \pi r^3 =$ $2/3 \, \pi r^3$. The volume of a sphere is twice this.

2.6.2 Circles and spheres

Any intersection between a plane and a sphere is a circle. Two well known intersections are lines of latitude and longitude.

A circle on the surface of a sphere is the plane that defines it passes through the center of the sphere. Lines of longitude are great circles because they go through both the North and South poles. The equator is a great circle, but other lines of latitude are not.

The shortest way to travel between two points on the surface of a sphere is to move along a great circle. This means that if you start at a point that is at 30°N latitude on the east coast of the United States and want to travel to a point that is at 30°N latitude on the west coast you should travel north of the 30th parallel.

2.6.3 Spherical triangle

I know I've told you many times in these notes that the angles in a triangle add up to 180°, but it turns out that this is not actually true. It would be true if the world were flat, but it isn't.

For example, suppose point A is the North Pole, point B is the point on the equator at 0° longitude, and point C is the point on the equator at 90° W longitude. It's pretty obvious that angles ABC and ACB are both right angles: they're formed by the intersection of a line of longitude and a line of latitude. If you think about what a globe looks like from the top, you'll realize that Angle BAC is also a right angle: it's formed by the intersection of two lines of longitude that are 90° apart. Hence, the sum of the angles in this spherical triangle is 270 degrees.

It's not true that all spherical triangles have angles that sum to 270 degrees. If you think about it, the triangles you're used to drawing are all spherical triangles (at least if you spread your paper out on the ground) and in elementary school you probably used protractors to verify that the angles did add up to about 180 degrees. The key difference is the size of the triangles. Yours were really small (compared to the Earth). Mine was not.

An interesting formula for the sum of the angles in a spherical formula is:

- Let ABC be a spherical triangle on the surface of a sphere of radius r. Let S be the sum of the measures of its angles and let A be its area. Then,

$$A = \frac{S - 180}{180} \pi r^2.$$

The triangles you're used to drawing have an area that's much smaller than the square of the radius of the earth. Hence, the sum of their angles must be very close to 180.

Note that the formula for the surface area of a sphere is also a corollary of this theorem. The triangle I described above covers one-eighth of the surface of the sphere. Its angles sum to 270°. Hence, the formula says that its area is ½ π r^2, which is one-eighth of what I'd told you the surface are of the sphere was.

2.6.4 Oblique cones and cylinders, frustums

An oblique cone lopsided cone in which the vertex is not directly above the center of the base. The V = ⅓ base × height formula holds for oblique cones (or pyramids). V = base × height also works for oblique cylinders.

If you cut off the pointy end of a cone the part that's left over is called a *frustum*. You can find the volume (or surface area) of a frustum by subtracting the volume of the cone you cut off the top from what the volume of the full cone was before you cut the top off.

Category 3 – Number Theory

The number theory category in meet #5 isn't about number theory. Instead, it is about set theory and Venn diagrams.

3.1 Venn Diagrams

A Venn Diagram is a convenient graphical way to keep track of two or three sets.

The diagram below is an example. It has three circles. The lower left circle represents the set A of sixth graders. The lower right circle represents the set B of people in the orchestra. The top circle represents the set C of girls. Anyone in the Bigelow math team (or not on the math team) can be put somewhere on this diagram to show which sets they belong to:

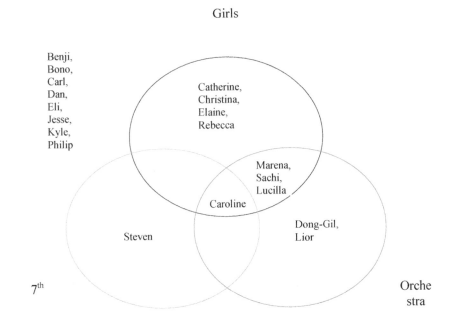

3.1.1 Using Venn Diagrams

The most common kind of Venn Diagram problem seems to be giving you a large number of facts and then asking you how many elements are in some set. For example,

Question (IMLEM Meet #5, Apr. 2005): Of the 125 members of the "Diver's Down" scuba club, 70 have been to the coral reefs, 80 have explored the shipwreck, and 60 have visited the underwater caves. Thirty-eight have been to both the coral reefs and the shipwreck, 42 have been to both the shipwreck and the caves, and 34 have been to the coral reefs and the caves. Only 23 divers have been to all three sites. How many members of the "Divers Down" scuba club have not been to any of these sites?

The easiest way to do problems like this is to use a Venn Diagram *and work backwards*.

1. Draw a Venn Diagram like the one on the next page.

2. We are told 23 divers have been to all three sites. This means there are 23 members in the area I've labeled c. Write 23 in this box to remind yourself that c=23.

3. The fact before this is that 34 divers have been to the reefs and the caves. This means that b+c=34. Hence, b=11. Write this down.

4. 42 divers have been to the shipwreck and the caves. This means c+f=42. Hence, f=19.

5. 38 divers have been to the reefs and the shipwreck. This means c+d=38. Hence d=15.

6. 60 divers have visited the caves. This means e+b+c+f=60. We know b=11, c=23, and f=19. Hence, e=7.

7. 80 have explored the shipwreck. This gives g+c+d+f=80. Hence, g=23.

8. 70 have been to the reefs. This gives a+b+c+d=70. Hence, a=21.

9. Finally, there are 125 divers. This tells us that a+b+c+d+e+f+g+h=125. From the above results we can compute a+b+c+d+e+f+g is 119. Hence, h=6, i.e. there are six divers left over who have not been to any of the three places.

Once you get used to these problems they go much faster. For example, you didn't need to do the last addition from scratch. You could have started instead from a+b+c+d=70.

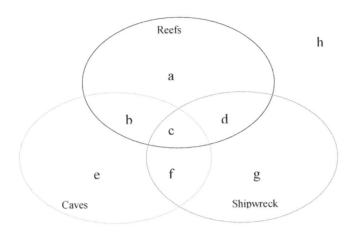

3.2 Basic Set Theory

3.2.1 Definitions

A *set* is a collection of elements. The standard notation for writing that A is the set containing elements a, b, and c is to write A = {a, b, c}. Note that the order doesn't matter in this notation. {c, a, b} and {a, b, c} are the same set.

The *union* of A and B, A $\cup$ B, is the set of all elements in A or B. For example, {1, 4}$\cup${2, 5} = {1, 2, 4, 5}.

The *intersection* of A and B, A $\cap$ B, is the set of all elements that are in both A and B. For example, {1, 2, 3}$\cap${1, 4, 9} = {1}.

The *empty set* is the set with no elements. It is written as {} or $\varnothing$. For example, {1, 2}$\cap${3}= $\varnothing$.

The *number of elements* of A is denoted by |A|. For example, if A={1, 4, 7, 10}, then |A|=4.

A set A is said to be a *subset* of B if every element of A is also an element of B. In this case we write A $\subseteq$ B. For example, {1}$\subseteq${1, 4, 9}, but {1, 5} is not a subset of {1, 4, 9}. Every set is a subset of itself. The empty set is a subset of any other set.

- If |A|=n then A has 2^n subsets.

3.2.2 Set theory calculations

Sometimes one can simplify set theory calculations using associative and distributive rules:

- $(A \cup B) \cup C = A \cup (B \cup C)$

- $(A \cap B) \cap C = A \cap (B \cap C)$

- $(A \cup B) \cap C = (A \cap C) \cup (B \cap C)$

The last of these comes in handy over and over again in meets in problems like the following:

Question: Suppose A is the set of natural numbers that are 2 more than a multiple of 7.
Suppose B is the set of natural numbers that are 5 more than a multiple of 17.
Suppose C is the set of natural numbers that are less than 40.
How many elements are in the set $(A \cup B) \cap C$?

It could take you a while to figure this out if you started by trying to figure out what $A \cup B$ looks like. If you instead start by writing $(A \cup B) \cap C = (A \cap C) \cup (B \cap C)$, then it's easy to quickly calculate $A \cap C = \{2, 9, 16, 23, 30, 37\}$ and $B \cap C = \{5, 22, 39\}$. The answer is 9.

One other tip for calculating expressions with intersections efficiently is to always try to start with the smaller set. For example, if you're asked

Question: Suppose A is the set of factors of 105.
Suppose B is the set of natural numbers less than 77 that are relatively prime to 77.
Find $| A \cap B|$.

In this case, because the prime factorization of 105 is $3 \times 5 \times 7$, you know that $A = \{1, 3, 5, 7, 3 \times 5=15, 3 \times 7=21, 5 \times 7=35, 3 \times 5 \times 7=105\}$. 1, 3, 5, and 15 are relatively prime to 77. If you'd started by trying to compute all numbers less than 77 that were relatively prime to 77 it would have taken a long time.

3.3 Inclusion-Exclusion Principle

A number of problems ask you to count how many elements are in a union of two or three sets. A very useful tool for doing this is the inclusion-exclusion principle. I'll write it out separately for the two set and three set cases. The best way to understand why this is true is to stare at a Venn Diagram.

- $|A \cup B| = |A| + |B| - |A \cap B|$.

- $|A \cup B \cup C| = |A| + |B| + |C| - |A \cap B| - |A \cap C| - |B \cap C| + | A \cap B \cap C|$.

You're first reaction may be that these long formulas could not possibly make it easier to compute the number of elements in a union of two or three sets. If so, you're wrong. Consider, for example, the following problem.

Suppose A is the set of natural numbers that are multiples of 3.
Suppose B is the set of natural numbers that are multiple of 5.
Suppose C is the set of natural numbers that are less than 100.
How many elements are in the set (A ∪ B) ∩ C?

The first step is to rewrite the problem as asking for $|(A \cap C) \cup (B \cap C)|$. The inclusion-exclusion principle says this is equal to $|A \cap C| + |B \cap C| - |A \cap B \cap C|$.

$A \cap C$ is just the set of numbers less than 100 that are multiples of 3. Hence, $|A \cap C| = 33$.
$B \cap C$ is just the set of numbers less than 100 that are multiples of 5. Hence, $|B \cap C| = 19$.
$A \cap B \cap C$ is the set of numbers less than 100 that are multiples of 3 and 5. This is the set of multiples of 15. Hence, $|A \cap B \cap C| = 6$.

The answer to the problem is $33 + 19 - 6 = 46$.

Here's another nice problem for practice:

Question: How many positive integers less than 1000 have a 7 in them? For example,
171 and 277 do and 45 does not.

One could do it by brute force, but it gets very confusing. To do it by inclusion exclusion, you let *A* be the set of numbers with a seven as the hundreds place, *B* be the set of numbers with a seven in the tens place, and *C* be the set of numbers with a seven in the ones place. Can you see how to get 271 as the answer now?

Some problems that sound like Venn Diagram problems are actually inclusion-exclusion problems in disguise. For example,

Question: There are 520 students in Bigelow Middle School. Forty-eight students play
in the orchestra. Eighteen are on the Math Team. Sixty-four are in the school musical.
Twelve students do exactly two of these activities. Four students do all three. How
many Bigelow students do none of the activities?

If you try to fill in a Venn Diagram you'll run into a problem: there's not enough information to fill in every cell. There is, however, enough information to answer the question. Let A be the set of students in the orchestra. Let B be the set of students on the math team. Let C be the set of students in the orchestra.

We know $|A \cup B \cup C| = |A| + |B| + |C| - |A \cap B| - |A \cap C| - |B \cap C| + |A \cap B \cap C|$. The first three sentences of the problem tell us that $|A| + |B| + |C| = 48 + 18 + 64 = 130$. The last sentence tells us that $|A \cap B \cap C| = 4$. The next to last sentence is a little trickier: the number of students who do exactly two activities is $|A \cap B| + |A \cap C| + |B \cap C| - 3|A \cap B \cap C|$, so it implies that $|A \cap B| + |A \cap C| + |B \cap C| = 12 + 3 \times 4 = 24$. Hence, $|A \cup B \cup C| = 130 - 24 + 4 = 110$. The answer is $520 - 110 = 410$.

Some other problems could be done as Venn Diagram problems, but are easier to do as inclusion exclusion problems. For example, suppose that in the above problem they gave you a few more facts:

Question: There are 520 students in Bigelow Middle School. Forty-eight students play in the orchestra. Eighteen are on the Math Team. Sixty-four are in the school musical. Eight students are in the orchestra and on the math team. Four students are on the math team and in the musical. Twelve students are in the orchestra and in the musical. Four students do all three. How many Bigelow students do none of the activities?

This problem gives you enough information to fill in the boxes so you could solve it that way, but it's quicker not to. Inclusion-exclusion says that the number of students doing at least one activity is $48 + 18 + 64 - (8 + 4 + 12) + 4 = 110$. The answer is still 410.

3.4 Review of Modular Arithmetic

For a number of past problems it was useful to have not forgotten how modular arithmetic works.

In modulo N arithmetic (also called mod N) the numbers are 0, 1, 2, 3, …, N-1, 0, 1, …, N-1, 0, 1, 2, … People sometimes picture this by thinking of numbers as lying on a number circle instead of on a number line.

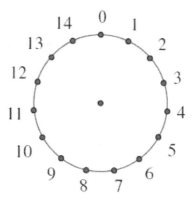

An alternate way to think about modular arithmetic is in terms of remainders:

- If r is the remainder when y is divided by x, then $y=r$ (mod x).

All of the formulas you use for adding, subtracting, and multiplying regular numbers still work in modulo N arithmetic. The only thing to watch out for is that numbers have multiple names. For example, -2=13 (mod 15) and 100=4 (mod 8). Examples of things you can still do in modular arithmetic are:

- $a + b = b + a$ (mod N)

- $a (b + c) = ab + ac$ (mod N)

- $(ab) c = a (bc)$ (mod N)

Modular arithmetic comes in handy in doing problems like problems like:

Suppose A is the set of natural numbers that are 3 more than a multiple of 8.
Suppose B is the set of natural numbers that are 1 more than a multiple of 7.
Suppose C is the set of natural numbers that are less than 500.
How many elements are in the set (A $\cap$ B) $\cap$ C?

Step 1. The first step in solving a problem like this is to find one number with these properties.

Here's how I'd do this. The numbers that are 3 more than a multiple of 8 are numbers of the form $8n + 3$ for n a nonnegative integer. Thinking in the mod 7 world I know that $8 = 1$ (mod 7). Hence, $8n + 3 = n + 3$ (mod 7). The smallest value of n that makes this equal to 1 is $n = 5$. This tells us that one number that is both 3 more than a multiple of 8 and 1 more than a multiple of 7 is 43.

Step 2. Remember that all elements of $A \cap B$ are of the form $43 + LCM(8,7) n$.

The least common multiple of 7 and 8 is 56, so this means that   is the set of numbers of the form $56n + 43$. These numbers are between 0 and 500 for n=1, 2, ... , 8. The answer to the problem is 8.

To see why all other solutions must be of the form $56n + 43$, note that if x is a solution to $x = 3$ (mod 8), then $x - 43 = 0$ (mod 8), so $x - 43$ must be a multiple of 8. Similarly $x = 1$ (mod 7), implies that $x - 43$ must be a multiple of 7. Any number that is a multiple of 7 and a multiple of 8 is a multiple of their least common multiple, 56.

To see if you understand this here's a similar challenge problem. If you can do this one you'll be all set for category 3.

Suppose A is the set of natural numbers that are 2 more than a multiple of 5.
Suppose B is the set of natural numbers that are 1 more than a multiple of 8.
Suppose C is the set of natural numbers that are multiples of 17.
Suppose D is the set of natural numbers that are less than 1000.
How many elements are in the set (A $\cap$ B $\cap$ C) $\cap$ D?

3.5 Advanced Topic: Venn Diagrams and Inclusion-Exclusion with Four or More Sets

3.5.1 Inclusion-Exclusion

The inclusion-exclusion principle works just as well for four or more sets as it does for three. For example, one can count the number of elements in the union of four sets by

$$|A \cup B \cup C \cup D| = |A| + |B| + |C| + |D|$$
$$- |A \cap B| - |A \cap C| - |A \cap D| - |B \cap C| - |B \cap D| - |C \cap D|$$
$$+ |A \cap B \cap C| + |A \cap B \cap D| + |A \cap C \cap D| + |B \cap C \cap D|$$
$$- |A \cap B \cap C \cap D|.$$

An example of a problem where you would use something like this is:

Question: How many natural numbers less than 100 have no prime factors that start with a 1?

You can think of this question as asking for $99 - |A \cup B \cup C \cup D|$, where A, B, C, and D are the set of numbers less than 100 that are a multiples of 11, 13, 17, and 19, respectively. The inclusion-exclusion counting is particularly easy here, because all sets but the first four are empty, e.g. $|A \cap B| = 0$ because the smallest number that is both a multiple of 11 and a multiple of 13 is 143. The answer is $99 - (9 + 7 + 5 + 5) = 73$.

The problem above was easy to count using inclusion-exclusion because so many of the sets were empty. This is not uncommon in math meet problems: a question writer would probably think it was unfair to ask a question where you needed to compute and add up 15 different terms to get the answer.

Another type of many set problem in which inclusion-exclusion counting is not overly time consuming is symmetric problems. For example, the 2007 meet asked multiple questions about rearranging drawers in a dresser. One was a word problem that essentially amounted to:

Question: How many ways are there to arrange the drawers in a dresser with four identical drawers so that at least one drawer is in its correct place?

Again, a good way to think about this problem is using inclusion-exclusion: the question is asking you to compute $|A \cup B \cup C \cup D|$, where A is the number of arrangements of four drawers in which the first drawer is in the right place, B is the number with the second drawer in the right place, and so on. The counting is not so hard in this one because of the symmetry. For example, we know $|A \cap B| = |A \cap C| = \ldots = |C \cap D|$. Hence, what's left to compute is

$$4\,|A| - 6\,|A \cap B| + 4|A \cap B \cap C| - |A \cap B \cap C \cap D|.$$

Most of these are pretty easy to compute (if you've read the part on category 4 already; otherwise don't worry about this). For example, $|A| = 6$ because after you've put the first drawer in the right place you can put the other three drawers in six different ways, and $|A \cap B \cap C| = 1$ because after you've put the first three drawers in the correct places there's only one place to put the last drawer. The answer to the question is 15.

If you've read section 4.1.3 the coefficients in above equation (4, 6, 4, and 1) may also look familiar: they're $_4C_1$, $_4C_2$, $_4C_3$, and $_4C_4$. You should be able to figure out why this is.

3.5.2 Venn diagrams

While many middle school classes cover Venn diagrams for two or three sets, I don't think I've ever seen one discuss how to draw a Venn diagram for four sets. Most kids probably assume this is because you can't do it on a two-dimensional piece of paper. If you're one of them, you're wrong. One can draw four set Venn diagrams on a piece of paper in many different ways. None is so simple as to make this a worthwhile way to do a problem that doesn't say you have to use a Venn diagram, but they are kind of neat.

The 2007 meet contains one example:

3. Venn diagrams work well for two or three sets, but not as well for four or more sets. The noodle-shaped region below shows one way that a fourth set might be included in a Venn diagram. If the natural numbers 1 through 72 inclusive are placed in the appropriate regions, what is the sum of the numbers in regions A, B, and C?

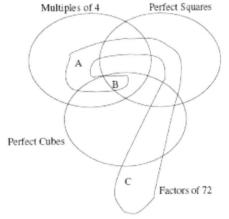

Answers
1. _____
2. _____
3. _____

The way to do this problem is just to not be freaked out by the diagram. Note that A, B, and C are subsets of the set of factors of 72. Hence, all you need to do is to consider the set S = {1, 2, 4, 8, 3, 6, 12, 24, 9, 18, 36, 72} of factors of 72 and figure out which of the other three sets they belong to. It turns out that 12, 24, and 72 are in A, B is empty, and C contains 2, 3, 6, and 18. The answer is 137.

The diagram above may make you think that you were wrong, but only on a technicality. Perhaps what you'd really thought was just that you couldn't draw a nice symmetric Venn diagram with four or more ordinary-looking sets. If so, you're still wrong. The diagram below shows how to draw a much prettier Venn diagram for a five set problem. It turns out that you can draw a symmetric Venn diagram in the plane for any prime number of sets. This is a hard theorem that was only proved in 2003.

The five set Venn diagram shown below was first discovered by Branko Grünbaum in the 1970s. The drawing is reproduced by permission of Frank Ruskey and Mark Weston. It and many other neat diagrams may be found in their "Survey of Venn Diagrams," published in the Electronic Journal of Combinatorics.

Category 4 – Arithmetic

The arithmetic category in meet #5 covers combinatorics and probability. I alternate between thinking it's a great category and a terrible category. The great part is that there's lots and lots of interesting math on it. The terrible part is that there's so much on it that most students will give up before getting to the end of these notes.

Some of the team questions on this topic have been extremely hard. The regular questions haven't been so bad.

4.1 Combinatorics

Combinatorics questions ask you how many ways something can be done. Most can be done quickly if you recognize whether they are asking you to compute a number of permutations or a number of combinations.

4.1.1 Orderings

A basic building block for many other counting problems is counting orderings. For example,

Question: A preschool class has 6 students. Every day they line up to go to lunch. They like to line up in a different order each time. After how many days will it no longer be possible to choose an order that they have not used before?

The way to think about this problem is that you can choose any of the six kids to be the line leader. After you've chosen the line leader, you have five choices for whom to put second. After choosing who goes first and second, you will have four kids to choose from when you're deciding who to put third, and so on. The total number of orderings is $6 \times 5 \times 4 \times 3 \times 2 \times 1 = 720$.

This is a big number! A school year typically has 180 days, so the class could make it through a year of preschool, a year of kindergarten, and first and second grade before running out of different orders to try. And this is just for a six student class.

A more important general fact to take away is

- The number of distinct orderings of an N element set is $N! = 1 \times 2 \times 3 \times \ldots \times N$.

If you're not already familiar with it, be sure to note the that the ! in this equation is a "factorial" sign. It means just what I wrote. For example, $3! = 6$, $5! = 120$, and $7! = 5040$.

4.1.2 Permutations

Permutations problems are like ordering problems, but with one difference: they ask you to count the number of ways you can order just k members out of a set of N elements. For example,

Question: Newton IMLEM meets used to have six teams: Bigelow, Brown-Blue, Brown-White, Day, Heath, and Oak Hill. The organizers want to unveil an engraved plaque at the end of the last meet giving the places of the first three teams. For example, it might say: 1st place: Bigelow; 2nd place: Day; 3rd place: Brown-Blue. (Assume that ties are not possible) They won't have time to get the plaque engraved between the end of the last meet and the ceremony, so they decide to get plaques engraved in advance listing the teams in every possible order and then throw out all the incorrect plaques after the meet. How many plaques do they throw out?

Question: At the start of the year a teacher buys one copy each of four DVDs (Star Wars, Shrek, Toy Story, and Harry Potter and the Sorcerer's Stone) and announces that she will give them out as prizes to the four students with the best attendance. At the end of the year, six students have perfect attendance records. In how many ways could she give out the DVDs?

The number of ways to choose k elements out of a set of N elements with order counting is known as the number of *permutations* and written $_NP_k$. The first question asks you for $_6P_3 - 1$. The second asks for $_6P_4$.

To solve these problems you start out just like you did with counting orderings: there are six teams that could finish first. After you've picked one team to be first, there are five teams that could be second. Then there are four teams that could be third. The difference from counting orderings is that you get to stop here – the plaque only has three slots so once you've picked someone to be third it's all full. The answer is $6 \times 5 \times 4 - 1 = 119$.

The -1 in the above answer comes from the fact that they asked "how many plaques to they throw out" rather than "how many plaques do they need to make up." This is a common trick of problem designers to get students to make mistakes. They're counting on you to be so excited that you know how to do it and get so caught up in counting the permutations that you don't read all the way to the end or forget why you were counting permutations. There's just one way to avoid getting tripped up by this: always be sure to read the question carefully and to reread it when you go back at it at the end to check over your work.

The second one works just like the first. There are six kids she could choose to get *Star Wars*. The question implies (although it doesn't say it directly) that no kid will get two DVDs. Hence, there are now five kids she can pick to get *Toy Story*. And so on. This gives $_6P_4 = 6 \times 5 \times 4 \times 3 = 360$.

I think it's instructive to do the problem in a different way too. This way is to think of any permutation problem as an ordering problem. To give out the DVDs I can choose a random ordering for the six kids and then give *Star Wars* to the first kid, *Toy Story* to the second, and so on. There are 6! orderings, so we've done this in 6! ways. The problem, however, is that these are not 6! *different* ways to give out the DVD. For example, the ordering Alice, Bob, Charlie,

David, Edd, and Frank, results in exactly the same allocation of prizes as the ordering Alice, Bob, Charlie, David, Frank, and Edd. Hence, counting the orderings is double-counting each prize allocation. The answer is $6!/2 = 720/2 = 360$.

You can think of any permutation problem this way. For example, if there were seven students instead of six with perfect attendance records, you'd start by saying that there are $7!$ possible orderings. When we count orderings instead of permutations, we're now doing an even more severe multi-counting. In this case an ordering that ends Edd, Frank, Glenn produces exactly the same allocation as an ordering that ends Edd, Glenn, Frank, and the same as Frank, Edd, Glenn, and the same as Frank, Glenn, Edd, and the same as Glenn, Edd, Frank, and the same as Glenn, Frank, Edd. Hence, we've counted each allocation six times as often as we should. The six here comes from the fact that there are $7 - 4$ people left over and we can order 3 people in $3! = 6$ ways.

The general formula is

- $$_NP_k = \frac{N!}{(N-k)!} = N(N-1) \dots (N-k+1).$$

Think about it in whichever way you find easier: (1) as the ordering problem with a denominator that comes from another ordering problem to account for the multi-counting; or (2) as a sequence of multiplications for each place on the ordered list. The second works version of a formula works better on a calculator, because $N!$ can be so big that the calculator starts rounding off and gets a wrong answer.

Some calculators have buttons labeled $_NP_k$. There is no rule against using such calculators in meet #5. so you might want to get one.

4.1.3 Combinations

Combinations problems ask you to find the number of groups of size k that can be formed from a set of N elements. For example,

Question: A group of 8 astronauts has been training for a mission to Mars. Only 5 of them will actually get to go to Mars. How many different groups of 5 astronauts can be chosen from the 8 astronauts?

Question: Fourteen students on the Bigelow math team will attend Meet #5. Ten students will be "official" and four students will be alternates. How many different ways can the official team be chosen?

The key difference between combinations problems and permutations problems is that in combinations problems the order in which people or objects are chosen to be in the group does not matter. For example, it makes no difference whether you write Caroline's name first on the sheet of official competitors and then put Christina's name second, or vice versa. The official math team is the same regardless of what order you list them on the sheet.

This is different from permutation problems in which Bigelow 1st and Day 2nd was very different from Day 1st and Bigelow 2nd.

The number of groups of size k that can be formed from a set of N elements is informally described as "N choose k" written $_NC_k$ or $\binom{N}{k}$. The first question asks you for $_8C_5$. The second asks for $_{14}C_{10}$.

There is a basically just one formula you need to know about combinations:

- $_NC_k = \dfrac{N!}{k!(N-k)!} = N(N-1)\ldots(N-k+1)/k!.$

The first version of the formula is probably the easier one to remember.

We can think of this in two different ways. Both involve thinking about double counting.

The first is to think of choosing a set by first choosing a permutation, e.g. we write down someone in the top space of the sheet of official people, then someone in the second slot, and so on. The number of ways to do this is $_{14}P_4 = 14!/4!$. These permutations, however, are not all different math teams. Any ordering of the same ten names is the same team. Hence, we've counted every team 10! times. This gives $_{14}C_4 = {}_{14}P_4 / 10! = 14! / (4!10!)$.

The second is to think of choosing a set b first choosing an ordering, e.g. write down the a complete list consisting of the person with the highest score on the practice test, then the person with the second highest, and so on down to the person with the lowest score, and then choose the team to be the first ten people on this list. The number of orderings is 14!. There's even more double counting going on here, however, because any reordering of the top ten *and* any reordering of the bottom four would product exactly the same team. Hence, the number of times we've counted each distinct team is 10! × 4!, and again the answer is 14! / (4!10!).

Again, you need to be careful when doing factorials on a calculator. The numbers often get so big that the calculator will start rounding them off and then get a wrong answer when it does a division. The second formula is better than the first in this regard, and can be made even better by canceling common terms. For example, the first formula says that answer to the math team problem is 14!/ (4! 10!). Note that the numerator of this fraction is *1 × 2 × 3 × … × 10 × 11 × 12 × 13 × 14* and the second term in the denominator is *1 × 2 × 3 × … × 10*. The second formula says that you can cancel the 10! terms and find that the answer is *(11 × 12 × 13 × 14) /(1 × 2 × 3 × 4)* = *11 × 13 × 7* = 1001. The canceling trick that I mentioned is that you can often find other terms that cancel, e.g. 12 = 3 × 4. Using this we find *(11 × 12 × 13 × 14) /(1 × 2 × 3 × 4)* = *11 × 13 × 7* = 1001.

Again, this is even easier if you have a calculator with a combinations button. If you do, you're calculator probably also has a factorial button. If it does, type on 0.5 and then press the factorial button. On many calculators this will produce an error message or the number zero. But others will return a number that starts with 0.886274. If you multiply this number by 2 and square it you'll find that it's telling you that ½! = $\sqrt{\pi}/2$. Why this is so is beyond the scope of this book. Perhaps some day I'll write a *Hard Math for High School*. I doubt it, however, so if you want to learn why you should probably just go on the Internet and do a search for the "gamma function."

4.1.4 Choice with replacement

In combinations and permutations problems you are only allowed to choose each element of the set at most once. For example, Bigelow could come in 1st or 2nd, but it cannot come in 1st and 2nd. In choice-with-replacement problems the same member of the set can be chosen multiple times.

For example, the teacher could give each the four DVDs at random and not worry about the fact that one student might get two or more while some others get none. Or the math team question could say that what will be engraved on the plaque is the school of the 1st place sixth grader, the school of the 1st place seventh grader, and the schools the school of the 1st place eighth grader.

Choice-with-replacement problems are even easier.

- The number of ways to choose k elements out of a set of N elements with replacement is N^k.

For example, there are $7^4 = 2401$ ways to give out four DVDs to seven potential recipients and $6^3 = 216$ possible plaques saying things like Top Sixth Grader: Day, Top Seventh Grader: Heath, Top Eighth Grader: Day.

To remember this, just think about the fact that there are seven people to give the first DVD to, then seven to give the second DVD to, then seven to give the third DVD to, then seven to give the fourth DVD to.

4.1.5 Counting subsets

Some combination-like problems don't specify the size of the group you need to choose. Consider, for example,

Question: A teacher has 20 students in her class. She needs to send at least one student down to the office to pick up something for her. How many ways can she choose a group of students to go down the office?

This question would be a combinations questions if they told you how many students had to be in the group. But they don't. Instead, you should think of this question as similar to:

Question: How many subsets does a set of size 20 have?

179

The answer to the subset question is 2^{20}. The way to see this is that you can think of choosing a subset as making 20 decisions. Is the first student in or out? Is the second student in or out? Is the third student in or out? And so on. If you have two choices to make at 20 different points in time there are 2^{20} different ways to choose in total.

The logic of this counting problem is similar to that of the counting-with-replacement problems.

Again, the first problem I gave above is one where reading and remembering the question is important. The answer is $2^{20} - 1$ instead of 2^{20}, because the empty subset is not a valid choice for the teacher.

4.2 Basic Probability

4.2.1 One randomizing device

In the simplest probability problems there are N equally likely things that can happen. For example, if a standard die is rolled the number that comes up can be 1, 2, 3, 4, 5, or 6 and all of these are equally likely.

In the simplest problems, you are asked how likely some *event* is, and you can figure out the answer by just figuring out in how many of the possible outcomes the event is true. For example,

Question: A six-sided die is rolled. What is the probability that the number on the top is prime?

A good way to start on any probability problem is to write down the equally likely outcomes:

$$1, 2, 3, 4, 5, 6$$

Then, cross out the ones for which the event (the number is prime) is not true and/or circle the ones where it is true. Here, three of the six numbers are prime (2, 3, and 5), so the answer to this problem is $3/6 = 1/2$.

4.2.2 Two randomizing devices

When doing basic probability problems you need to *be very careful to be sure that the set of outcomes you wrote down are all equally likely*. The most common example where you could do this wrong are problems that involve two randomizing devices. For example, suppose you are asked:

Question: A standard six-sided die is rolled twice. What is the probability that the sum of the two numbers is 10?

You might think that you should start by saying that there are 11 possible outcomes – the sum could be 2, 3, 4, …, 12. Starting this way would be a bad idea. These 11 outcomes are **not** equally likely, so the answer to the problem is **not** 1/11.

Instead, the way you start is to say that there are 36 equally likely outcomes: a 1 followed by a 1; a one followed by a 2; a 1 followed by a 3; etc. It's sometimes helpful to think of these as forming a big grid:

$$
\begin{array}{cccccc}
(1,1) & (1,2) & (1,3) & (1,4) & (1,5) & (1,6) \\
(2,1) & (2,2) & (2,3) & (2,4) & (2,5) & (2,6) \\
(3,1) & (3,2) & (3,3) & (3,4) & (3,5) & (3,6) \\
(4,1) & (4,2) & (4,3) & (4,4) & (4,5) & (4,6) \\
(5,1) & (5,2) & (5,3) & (5,4) & (5,5) & (5,6) \\
(6,1) & (6,2) & (6,3) & (6,4) & (6,5) & (6,6)
\end{array}
$$

If you look at this grid for a minute, you'll see that there are three outcomes with a sum of 10 in a diagonal line in the lower right part of the grid: circle (4,6), (5,5), and (6,4). This is 3 of the 36 outcomes, so the answer is 3/36 or 1/12.

An example from the 2003 meet is

Question (IMLEM Meet #5, Apr. 2003): An eight-sided die is rolled twice. What is the probability that the sum of the two numbers is 10 or more?

In this problem the grid is a little bigger.

$$
\begin{array}{cccccccc}
(1,1) & (1,2) & (1,3) & (1,4) & (1,5) & (1,6) & (1,7) & (1,8) \\
(2,1) & (2,2) & (2,3) & (2,4) & (2,5) & (2,6) & (2,7) & (2,8) \\
(3,1) & (3,2) & (3,3) & (3,4) & (3,5) & (3,6) & (3,7) & (3,8) \\
(4,1) & (4,2) & (4,3) & (4,4) & (4,5) & (4,6) & (4,7) & (4,8) \\
(5,1) & (5,2) & (5,3) & (5,4) & (5,5) & (5,6) & (5,7) & (5,8) \\
(6,1) & (6,2) & (6,3) & (6,4) & (6,5) & (6,6) & (6,7) & (6,8) \\
(7,1) & (7,2) & (7,3) & (7,4) & (7,5) & (7,6) & (7,7) & (7,8) \\
(8,1) & (8,2) & (8,3) & (8,4) & (8,5) & (8,6) & (8,7) & (8,8)
\end{array}
$$

The set of pairs that sum to at least 10 form a triangle on the lower right side. Circle them. To count how many pairs are circled notice that its one pair in the second row, two pairs in the third, …, and seven in the bottom row. 1+2+3+4+5+6+7=7 ×8 / 2 = 28. The grid has 64 pairs. The answer is 28/64 = 7/16.

In questions that ask about multiplying numbers it helps to think about prime factorizations.

Question (IMLEM Meet #5, Apr. 2001): A standard six-sided die is rolled twice. What is the probability that the product of the two numbers is a multiple of six?

Again, you start by drawing the grid of equally likely outcomes.

(1,1) (1,2) (1,3) (1,4) (1,5) (1,6)
(2,1) (2,2) (2,3) (2,4) (2,5) (2,6)
(3,1) (3,2) (3,3) (3,4) (3,5) (3,6)
(4,1) (4,2) (4,3) (4,4) (4,5) (4,6)
(5,1) (5,2) (5,3) (5,4) (5,5) (5,6)
(6,1) (6,2) (6,3) (6,4) (6,5) (6,6)

Obviously, all pairs in the bottom row and the rightmost column will work. Circle them. The other way to get a product that is divisible by six is to have one number be a multiple of 2 and the other be a multiple of 3. This gives four more pairs: (2,3), (4,3), (3,2), and (3,4). Circle them. In total, you've circled 15 pairs, so the answer is 15/36 = 5/12.

One tip for doing problems under time pressure during a meet is to not actually write in all the numbers in the grid. What I would do instead is to just put the possible values for the first and second die on the row and column headings and then circle spaces in the grid. To get comfortable doing this you could try to solve the three problems again drawing a table like the one below instead of filling in all the numbers in the grid.

	1	2	3	4	5	6
1						
2						
3						
4						
5						
6						

4.3 Probability Problems with Two Events

Many probability problems talk about two events and ask questions with "and" or "or" in them.

4.3.1 "And" problems with independent events

The easiest problems involving multiple events are problems where the two events are *independent*. Roughly speaking, two events are independent if knowing whether the first one has occurred tells you nothing about whether the probability that the second one will occur.

Question: A six-sided die is rolled and a twelve-sided die is rolled. What is the probability that the numbers on top of both dice are prime?

These event that "the number on top of the six-sided die is prime" and "the number on top of the twelve-sided die is prime" are a classic example of independence. What happens on one die has nothing to do with what happens on the other die.

In problems with independent events there is a simple formula for figuring out the probability of two events: you just multiply the two probabilities.

- Prob(Event1 and Event2) = Prob(Event1) × Prob(Event2)

Here, the probability that the number on top of the six-sided die is prime is 3/6=1/2. (The six sides are equally likely and three of them have prime numbers: 2, 3, 5.) The probability that the number on top of the twelve-sided die is prime is 5/12. (There are twelve equally likely numbers and 2, 3, 5, 7, and 11 are prime).

The probability that both numbers are prime is 1/2 × 5/12 = 5/24.

4.3.2 "And" problems with dependent events

More often in probability problems the two events are not independent. For example, in 2001 IMLEM asked:

Question (IMLEM Meet #5, Apr. 2001): There are six butterscotch and 8 peppermint candies in a jar. What is the probability of drawing first a butterscotch and then a peppermint candy from the jar?

Many students probably answered 6/14 × 8/14 = 12/49. *They were wrong!* The reason they were wrong is that the two events "the first candy drawn is butterscotch" and "the second candy drawn is peppermint" are NOT independent.

The way to compute probabilities in problems like this is to use *conditional probabilities*:

- Prob(Event1 and Event2) = Prob(Event1) × Prob(Event2 | Event1).

The notation Prob(Event2 | Event1) is read as "the probability of Event2 conditional on Event1". It means the probability that Event2 happens *given* that we know Event1 has already happened.

In the problem above, let

    Event1="the first candy drawn is butterscotch"
    Event2 = "the second candy drawn is peppermint"

The first thing we need to recognize is that

    Prob(Event1) = 6/14 = 3/7.

This is pretty obvious. All 14 candies are equally likely to be chosen. Six of them are butterscotch.

The second thing we need to figure out is that

Prob(Event2|Event1) = 8/13.

This is only a little harder to see. Once we know that the first candy drawn was butterscotch, we know that there were 13 candies left in the jar before the second one was drawn and 8 of those were peppermint. This makes the conditional probability 8/13.

The answer to this problem is $3/7 \times 8/13 = 24/91$.

A similar problem is

> *Question (IMLEM Meet #5, Apr. 2002): There are 5 red lollipops, 3 purple lollipops, and 4 green lollipops in a basket. Assuming that each lollipop is equally likely to be picked, what is the probability that the next three lollipops taken will be green?*

The answer to this problem is $4/12 \times 3/11 \times 2/10 = 1/55$.

In the MathCounts part of this book I discuss a number of problems involving choices from more complicated state spaces. Many of them could also be solved using conditional probabilities. If you want more practice, you could skip ahead and try them now. One of the nice things about math is that if you do problems correctly, you get the same answer regardless of which method you use.

4.3.3 "Or" problems

IMLEM seems to have asked many more "And" questions than "Or" questions, but I don't see why this should be. For example, they could have asked:

> *Question: A six-sided die is rolled and a twelve-sided die is rolled. What is the probability that at least one of the two numbers that comes up is prime?*

There are two useful tricks for doing "Or" problems.

Trick 1 – Inclusion-Exclusion: One is a formula related to the inclusion-exclusion formula I talked about in the set theory part:

- Prob(A or B) = Prob(A) + Prob(B) – Prob(A and B).

In the problem above, let

A = "The number on the six-sided die is prime."
B = "The number on the twelve-sided die is prime."

As above we have:

Prob(A) = 3/6= 1/2

Prob(B) = 5/12

Prob(A and B) = 1/2 ×5/12 = 5/24

Hence, the answer is

Prob(A or B) = 1/2 + 5/12 − 5/24 = 17/24.

<u>Note</u>: One easy special case is problems where "A and B" is impossible. In this case the inclusion exclusion formula becomes

- Prob(A or B) = Prob(A) + Prob(B)   IF it's impossible for both A and B to occur.

For example,

Question: A twelve sided die is rolled. What is the probability that the number that comes up is prime OR a perfect square.

No numbers are both prime and a perfect square. Hence, the answer is 5/12 + 3/12 = 2/3.

<u>Trick2 – Negation</u>: The other trick is to focus instead on the probability that the sentence is not true. This transforms an "Or" problem into an "And" problem. Here, the statement "the first number is prime OR the second number is prime" is false whenever the statement "The first number is not prime AND the second number is not prime is true".

The probability of the latter statement is an easy "And" problem. The probability that the number on the six-sided die is not prime is 1/2. The probability that the number on the twelve sided die is not prime is 7/12. The probability that the first is not prime and the second is not prime is 1/2 × 7/12 = 7/24.

This tells us that the probability that the statement "the first number is prime OR the second number is prime" is false is 7/24. The probability that it is true is thus 1 − 7/24 = 17/24.

In terms of algebra the second trick can be thought of as the formula:

- 1 − Prob(A Or B) = Prob(Not A And Not B).

One famous probability problem which you do using trick two is the birthday problem:

Question: Suppose there are thirty kids in a class. If each kid is equally likely to be born on each of the 365 days of the year (and there are no twins in the class), what is the probability that two or more kids share a birthday.

The way to attack this problem is to compute the probability that no two kids have the same birthday. This is a big calculation using conditional probabilities: it's the probability that the second kid's birthday is different from the first AND the third kid's is different from the first and second, AND the fourth kid's is different from each of the first three … The answer is

$$(364/365) \times (363/365) \times (362/365) \times (361/365) \times (360/365) \times \ldots \times (336/365) \approx 0.29.$$

(One thing to be careful about here is that there are only 29 terms not 30. The first thing you need is for the *second* kid's birthday to be different from the first.) Hence, there is a 71% chance that two kids will have the same birthday.

4.4 Advanced Topic: Using Conditional Probabilities for Complicated Events

Many probability problems ask about the probability of some complicated event. For example,

> *Question (IMLEM Meet #5, Apr. 2001): A standard six-sided die is rolled twice. What is the probability that the product of the two numbers is a multiple of six?*

> *Question (IMLEM Meet #5, Apr. 2003): On the Greedy Grab game show, a contestant is blindfolded and draws two bills from a bag. If the bag is always stocked with 1 one-hundred dollar bill, 2 fifty-dollar bills, 3 twenty dollar bills and 4 ten dollar bills. What is the probability that a contestant will draw at least 100 dollars in total?*

One can do each of these problems by drawing tables of all of the possible combinations just like we did in section 4.2.2.

One can also do them using conditional probabilities. The way to do this is to think of several mutually exclusive first events for which the probability of that event AND the conditional probability of the events they ask about are easy to figure out. For example, in the Greedy Grab Game we have

Prob(Total$\geq$100) = Prob(1st is 100 AND Total$\geq$100) + Prob(1st is 50 AND Total$\geq$100) + Prob(1st is 20 AND Total$\geq$100) + Prob(1st is 10 AND Total$\geq$100).

Prob(1st is 100 AND Total$\geq$100) = Prob(1st is 100)Prob(Total$\geq$100|1st is 100)=1/10$\times$1 = 1/10.
Prob(1st is 50 AND Total$\geq$100) = Prob(1st is 50)Prob(Total$\geq$100|1st 50)=2/10$\times$2/9 = 4/90.
Prob(1st is 20 AND Total$\geq$100) = Prob(1st is 20)Prob(Total$\geq$100|1st 20)=3/10$\times$1/9 = 3/90.
Prob(1st is 10 AND Total$\geq$100) = Prob(1st is 10)Prob(Total$\geq$100|1st 10)=4/10$\times$1/9 = 4/90.

The answer is 1/10 + 4/90 + 3/90 + 4/90 = 20/90 = 2/9.

Note: The original answer sheets gave an incorrect answer to this question.

4.5 Averages

Three facts about averages are useful for answering IMLEM questions:

- If the average of n numbers is a then the sum of the n numbers is $s = n \times a$.

- If the average of n_1 numbers is a_1 and the average of n_2 numbers is a_2 then the average of all $n_1 + n_2$ numbers is $(n_1 a_1 + n_2 a_2) / (n_1 + n_2) = (n_1/(n_1 + n_2)) a_1 + (n_2/(n_1 + n_2)) a_2$.

- If the average of n_1 numbers is a_1, the average of n_2 numbers is a_2, and the average of all $n_1 + n_2$ numbers is a, then $n_1 (a_1 - a) = - n_2 (a_2 - a)$.

A good way to think about the third fact is to recognize that the total amount by which some numbers are above average must exactly offset the amount by which the other numbers are below average. For example, suppose you've taken five quizzes so far in a class and your average is 93. If you're thinking about what you'd need to do to keep your average at 90 or better, you'd think that you've got 15 points to spare right now. That means that on your next quiz you could get 15 points below 90, i.e. get a 75, and still have a 90 average. If there are three quizzes to go you could use up the 15 point cushion by averaging 85 on the last 3 quizzes.

A typical question is:

Question (IMLEM Meet #5, Apr. 2001): Albert had an average of 88% on six quizzes, all of which were weighted equally. If 100% is the highest score he could have received on any of the quizzes, what is the lowest possible score he could have received on one of the quizzes.

The most foolproof way to answer any of these questions is to use the first fact. If his average on six quizzes was 88, then the sum of his scores was $6 \times 88 = 528$. The way to get this sum with the lowest score on one quiz is to have 100 on five quizzes and 28 on the other.

The third fact lets you do this in your head slightly more quickly. The greatest amount by which he could have scored above 88 on five good quizzes is $5 \times 12 = 60$. Hence, the lowest he could have done on the other quiz is $88 - 60 = 28$.

A couple examples of questions where the second and third rules make things easy are:

Question: 45% of the members of the Bigelow math team are girls. One-third of the girls prefer doing Mystery to Algebra. All of the boys prefer Mystery. What percent of the math team prefers Mystery?

Question: A chemistry teacher gave A's to exactly half of his students. He gave A's to two-thirds of the students in his "honors" classes. In his other classes 46 of the 114 students got A's. How many students are in his honors classes?

The first question is one you can immediately apply the averages of averages formula: the answer is $0.45 (1/3) + 0.55 (1) = 0.70 = 70\%$.

The second is a direct application of the third rule. Writing n_1 for the number of students in the honors class we have $n_1 (\frac{2}{3} - \frac{1}{2}) = 114 (\frac{1}{2} - 46/114)$ which gives $n_1 (1/6) = 57 - 46$. The answer is 66. Verbally, my first thought after reading this question was "Oh. He gave 46 of 114 students A's. Half would be 57 so he's 11 behind. This means that in his other class he must have given A's to 11 more than half of the students. $2/3 - 1/2 = 1/6$, so the 11 extra A's are 1/6 of the class. The answer must be 66."

Category 5 – Algebra

The algebra category in meet #5 covers quadratic equations.

5.1 Solving Quadratic Equations

A quadratic equation is an equation of the form $ax^2 + bx + c = 0$.

A quadratic equation can have zero, one or two real solutions (also called *roots*). You can tell by computing $b^2\text{-}4ac$. If this is positive there are two solutions. If it is zero there is one solution. If it is negative there are no solutions.

There are two main ways to find the solutions to quadratic equations.

5.1.1 Quadratic formula

One way that always works is to simply use the quadratic formula. The quadratic formula says that the two solutions to a quadratic equation are:

$$x = \frac{-b \pm \sqrt{b^2 - 4ac}}{2a}.$$

The $\pm$ symbol in this equation means that you get one solution by using the $-$ sign and the other by using the $+$ sign.

For example, if you're given the equation $x^2 - 5x + 6 = 0$ you just plug $a = 1$, $b = -5$, and $c = 6$ into the formula. Computing $b^2\text{-}4ac = 25 - 24 = 1$ first tells us that there are two solutions. They are $(5 - 1)/2$ and $(5 + 1)/2$, otherwise known as 2 and 3.

5.1.2 Factoring

The quadratic formula is pretty simple and always works. Nonetheless, you can sometimes solve quadratic equations even more quickly by *factoring*.

If you multiply $(x - r_1)$ by $(x - r_2)$ you get a quadratic that is zero at both r_1 and r_2. The factoring method of solving a quadratic is essentially to look at it and recognize the two factors that must have been multiplied together to get it.

To see how this works, note that $(x - r_1)(x - r_2) = x^2 - (r_1 + r_2)x + r_1 r_2$. Hence, if you're given a quadratic equation like $x^2 - 5x + 6 = 0$, you know the roots are two numbers that add up to 5 and have 6 as their product. The answer, of course, is 2 and 3.

189

The factoring method tends to works best when the solutions are integers and c is prime. For example, if you're given the equation $x^2 - 18x + 17 = 0$, you know that the product of the two roots is 17. The only way to get a product of 17 with integers is 1 and 17. They sum to 18, so they are the solution.

One thing to be careful of when factoring is to be careful about negative numbers. The roots of $x^2 + 18x + 17 = 0$ are –1 and –17. The roots of $x^2 - 16x - 17 = 0$ are –1 and 17. If you haven't had algebra before it's probably safer to stick with the quadratic formula. Of course, if you haven't had algebra before, hopefully you're just doing this category as an alternate and don't need to worry about being safe.

5.1.3 Sums and products of roots

Suppose that r_1 and r_2 are the roots of $ax^2 + bx + c = 0$ and that $r_1 < r_2$. Some nice formulas are:

- $r_1 + r_2 = -b/a$.

- $r_1 r_2 = c/a$.

- $r_2 - r_1 = \dfrac{\sqrt{b^2 - 4ac}}{a}$.

These sometimes let you find answers to questions without finding the roots themselves. One example is:

Question (IMLEM Meet #5, Apr. 2001): There are two solutions to the equation $x^2 + 6x = 91$. If these two solutions were plotted on the real number line, what would the distance between them be?

The question is asking for the (absolute value of the) difference between the two roots. The last formula tells you that this is $\sqrt{36 - 4 \cdot (-91)}/1 = 20$.

5.2 Graphs of Quadratic Equations

The graph of the quadratic equation $y = ax^2 + bx + c$ has a shape known as a *parabola*. Some facts about this graph are:

- If $a > 0$ the graph points up at the ends as in the figure below. If $a < 0$ the graph points down at the two ends (like an upside down version of the same figure).

- The places where the parabola intersects the x-axis are the roots of the quadratic equation $ax^2 + bx + c = 0$. This means that the quadratic in the picture below has –3 and –4 as its two roots.

- If the graph doesn't intersect the x-axis, then the quadratic has no real roots. If it intersects the x-axis at exactly one point r, then the equation must be $y = a (x - r)^2$.

- The parabola $y = ax^2 + bx + c$ intersects the y-axis at c. This means that the equation being graphed in the figure below has $c = 12$.

- If $a > 0$ then the minimum value of y occurs when $x = -b/2a$. If the parabola has two real roots, then the value of x for which the function is minimized is also the average of the two roots. The parabola below, for example, is at its minimum value when $x = -3.5$.

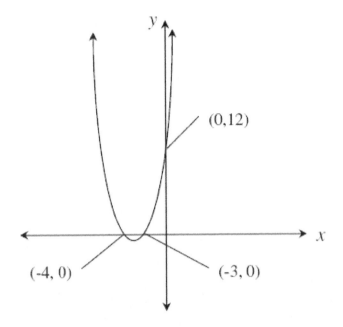

One thing you should feel comfortable doing is going from graphs to equations and vice-versa. For example,

> Question: *Find the equation for the parabola shown in the figure above.*

From the second fact we know that the roots are -3 and -4. The only quadratics with these roots are equations of the form $y = a (x + 3)(x + 4)$ for some value of a. The fourth fact then tells us that $a \times 3 \times 4 = 12$, which implies that $a=1$. The answer is $y = x^2 + 7x + 12$.

Here's a harder problem from 2007. I assume they intended to say that the equation was quadratic as opposed to "qaudratic." Qaudratic is not a word.

4. The graph of the quadratic equation shown at right crosses the x-axis at the points (-5,0) and (7,0), and its vertex is at the point (1,-36). Give the coordinates of the two points where the graph of this same equation crosses the line y=13. Reminder: The coordinates of the two points should be written as ordered pairs in parentheses.

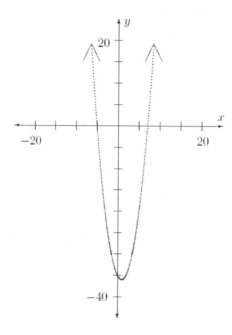

The standard way to do this problem would be to first find the equation. Given that the roots are -5 and 7, we know that the parabola must be of the form $y = a(x + 5)(x - 7)$. We know that $y = -36$ when $x = 1$, so $a = 1$. To finish the problem, we now need to find the solutions to $(x + 5)(x - 7) = 13$. You can do this in several ways: (1) guessing. If two whole numbers are going to multiply together to give 13 they must be 13 and 1 or -13 and -1. The product above will be 13 $\times$ 1 when $x = 8$. It will be -1 $\times$ -13 when $x = -6$; (2) solving the quadratic obtained by subtracting 13 from both sides: $x^2 - 2x - 35 = 13$ implies $x^2 - 2x - 48 = 0$. We can factor this or use the quadratic formula to find that the solutions are $x = -6$ and $x = 8$.

Note that this was a problem where you had to be careful to remember what they asked. The correct answer is (-6, 13) and (8, 13).

5.3 Problems That Involve Quadratic Equations

Many IMLEM problems are word problems that give a quadratic equation.

5.3.1 Problems with reciprocals

One source of quadratic equations is problems with reciprocals. These turn into quadratic equations when you multiply through to eliminate the denominator. For example,

Question: The sum of a number and three times its reciprocal is 4. What is the number?

The problem tells us that $x + 3(1/x) = 4$. Multiplying both sides by x. This gives $x^2 + 3 = 4x$. Subtracting $4x$ from each side gives a standard quadratic equation: $x^2 - 4x + 3 = 0$. The solutions are 1 and 3.

5.3.2 Sums and products of two numbers

Problems in which you are told the sum and product of two numbers are also solved using quadratic equations. Some problems like this come up in geometry.

Question (IMLEM Meet #5, Apr. 2001): A rectangular room has perimeter 60 feet and area 216 square feet. How many feet longer is the room than it is wide?

Let x and y be the length and width of the rectangle. Mathematically, the question tells us $2x + 2y = 60$ and $xy = 216$. One good way to solve a problem like this is to think of the factor pairs of 216 and see which add up to 30.

If you can't figure it out right away (or if they've chosen a problem where the solutions are not integers), you can always solve problems like this using quadratic equations.

The first equation tells us that $y = 30 - x$. Plugging this into the formula for the product gives
$$x(30 - x) = 216.$$
Multiplying this out gives
$$30x - x^2 = 216.$$
I like to keep the coefficient on the quadratic term positive so I bring everything over to the right side:
$$x^2 - 30x + 216 = 0.$$
The solutions are 12 and 18.

The problem I gave above is an example where you probably could have done the problem more quickly by just looking at factor pairs. If I'd told you that the area was 215, however, the solution to the problem would be that the sides of the rectangle are $15 - \sqrt{10}$ and $15 + \sqrt{10}$. This is something you wouldn't have found without using the quadratic formula.

5.3.3 Pythagorean Theorem

The Pythagorean Theorem is another common source of quadratic equations. For example,

Question (IMLEM Meet #5, Apr. 2002): Find the side length of a square that has the same area as the right triangle shown below. Round your answer to the nearest whole number of units.

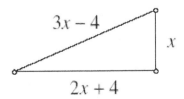

193

The first step in this problem is finding x. To do so you solve the Pythagorean relationship:
$$x^2 + (2x+4)^2 = (3x-4)^2.$$
This expands to
$$x^2 + (4x^2 + 16x + 16) = 9x^2 - 24x + 16.$$
Cancelling terms gives
$$0 = 4x^2 - 40x.$$
The solutions are $x = 0$ and $x = 10$. Only the latter is a valid length for a side in a triangle.

The area of the triangle is 120. The side length of the square is $\sqrt{120}$, which is very close to 11.

5.3.4 Problems with speeds and distances

Problems involving speeds and distances often end up giving quadratic equations. For example,

Question (IMLEM Meet #5, Apr. 2005): Rosie traveled 780 miles. If her average speed had been 5 miles per hour faster, the trip would have taken one hour less. What was her average speed?

Let x be Rosie's average speed in miles per hour. The time her trip took is $780/x$ hours. The problem tells us that $780/(x+5) = 780/x - 1$. Multiplying both sides by $x(x+5)$ gives

$$780x = 780 (x+5) - x (x+5).$$

Simplifying this we get the quadratic equation

$$x^2 + 5x - 3900 = 0.$$

The roots are -65 and 60. Her speed must be a positive number, so the answer is 60 miles per hour.

5.3.5 Quadratic equal to some number

A sample problem of this type is:

Question: For what value of x is $x^2 - 3x - 8$ equal to 100?

In equation form this problem is $x^2 - 3x - 8 = 100$. Subtract 100 from both sides to get $x^2 - 3x - 108 = 0$. The solutions are -9 and 12.

5.3.6 Finding one root given the other

A sample problem is:

Question: The roots of $x^2 - 5x = b$ are 14 and a. Find a and b.

- Method 1: In standard form the quadratic equation is $x^2 - 5x - b = 0$. From the result I mentioned on the sum of the roots, you know the sum of the two roots is 5. Hence $a = -9$. From the result on the product of the roots we know the product of roots is $-b$. The product of -9 and 14 is -126. Hence, we know $b = 126$.

- Method 2: Plugging in 14 for x in the equation gives $b = 14^2 - 5 \times 14 = 126$. Then solve $x^2 - 5x - 126 = 0$ to find $a = -9$.

5.4 Advanced Topic: Complex Numbers

If you don't know what a complex number is you should not read the rest of this section. If you do know what a complex number is you could read on if you like, though I haven't seen this come up on IMLEM meets.

When a quadratic equation has $b^2 - 4ac < 0$ I told you above that the quadratic has no real solutions. This is true. What I didn't tell you, however, was that in this case the quadratic has two solutions. It's just that the solutions are complex numbers instead of real numbers. For example, the quadratic equation $x^2 + 1 = 0$ has two solution i and $-i$.

Complex roots always come in what are called conjugate pairs: if one root is $f + g\,i$, then the other is $f - g\,i$.

There's really nothing else you need to know about quadratic equations with complex solutions. You can do everything just like you did before. You can use the quadratic formula to find the solutions. The equations for the sum, product, and difference between the roots all work. And so on.

A quadratic will have complex roots only if it has no real roots. Hence, every quadratic has either one or two roots (if you count both real and complex roots).

Both IMLEM and MathCounts seem to try to avoid using complex numbers, but occasionally they introduce them by accident. Consider for example,

Question: Suppose that $x + 1/x = 7/4$. Find $x^2 + 1/x^2$.

There's a neat trick for doing this problem: note that $(x + 1/x)(x + 1/x) = x^2 + 2 + 1/x^2$. Hence, $x^2 + 1/x^2 = (x + 1/x)^2 - 2 = 49/16 - 2 = 17/16$.

What does this have to do with complex numbers? It turns out that this problem is all about complex numbers. For any positive real number $x + 1/x$ is always at least 2. Hence, if there were only real numbers the answer to the problem would be that there is no such value of x. We can, however, easily solve the equation $x + 1/x = 7/4$ for complex solution. Multiplying both sides

195

by x and then using the quadratic formula we find that the solutions

are $x = \dfrac{7 \pm \sqrt{-15}}{8} = \dfrac{7}{8} \pm \dfrac{\sqrt{15}}{8} i$.

The fact that the answer to a math problem will be the same regardless of how you do it (as long as you do it right) tells us that it must be the case that

$\left(\dfrac{7}{8} + \dfrac{\sqrt{15}}{8} i \right)^2 + 1 / \left(\dfrac{7}{8} + \dfrac{\sqrt{15}}{8} i \right)^2 = 17/16$. This may seem unlikely, but if you multiply it out

you'll find that it really is true. (Simplifying the resulting expression is a nontrivial problem. You need to remember that the way to rationalize $\dfrac{1}{a + \sqrt{b}}$ is to multiply the numerator and denominator by $a - \sqrt{b}$.)

MathCounts®†

MathCounts® is the most popular middle school math contest in the United States. About 6000 schools participate.

Each MathCounts contest has a similar structure. First, there is a "Sprint round". It's a 30 question test that you only get 40 minutes to do. The questions are easier than IMLEM questions, but they give you much less time. Stupid mistakes kill you on the sprint round, so most kids will be better off setting a goal and being a little more careful not to screw up on the easiest questions rather than trying to do the last few. Second there is the "Target round". It is actually four separate rounds: on each you get 6 minutes to do two questions. Not much strategy here. Mostly just do the problems and try to get them right. On the last couple rounds you should do the easier question first because you may not have time to do both. Each target problem counts for two points, so the whole round is worth 16 points. This makes the maximum individual score 46 points. Finally, there is the team round. The team round is extremely important to the team score: each team question counts as much as eight sprint questions. The MathCounts team rounds are oddly unrushed. You get 20 minutes to do the ten problems. Usually only the last one or two questions are hard, although sometimes one or two of the early problems will be also be things where it's very easy to make a mistake. If you don't get 9 or 10 problems right, you'll lose ground relative to teams that do. The one important strategic aspect is to make sure you use your top student appropriately: ideally you'd like to have your best student spend most of his or her time getting numbers 9 and 10 right, but it's really, really important not to get the early problems wrong, so your other team members need get good at recognizing early on that they're not sure of an answer and should ask for help.

MathCounts starts with a "School" contest mailed out to participating schools in January. A very nice feature of the MathCounts program is that this contest can be freely copied and given to as many kids in a school as are interested in taking it (or forced to take it by their teachers). The school contest is purely for internal awards. No results are reported back to the MathCounts organization.

In February comes the "Chapter" contest. Each school picks a four-student team to represent it at a competition held at some nearby school. Schools can also bring up to four other students as "individual" competitors. They take the same tests and compete for the same individual prizes, but their scores don't count toward the team total. Some states divide up into many small chapters with just a handful of schools at each. Other states bring dozens of schools to each chapter site. The Metro Boston chapter meet has about two hundred students from thirty or so schools and is quite a sight.

---

† Mathcounts® is a registered trademark of the Mathcounts Foundation, which was not involved in the production of, and does not endorse, this book. Indeed, they disliked aspects of an earlier draft including that "several assumptions and generalizations which may not be accurate are stated as fact" and that my opinions "could be misinterpreted as actual stated program goals." I assume that they would feel similarly about the current draft. So it would probably be more informative to say that they not only do not endorse this chapter, but would prefer that it not be published.

The top few schools and top few individuals from each chapter contest advance to "State" competitions, which are usually held in March. In Metro Boston the top five or so schools advance along with the top five or so people among those who were competing as individuals or were on teams that did not advance. Other schools may also advance through a wild-card process. Massachusetts chapters with fewer participating schools get fewer automatic qualifiers. Some other states have similar procedures, but there's a lot of heterogeneity: in some small states every school gets to advance; in others you may need to win your chapter; in some even winning the chapter isn't a guarantee. Qualifying as a team is a big accomplishment. In Metro Boston, if your four team members average around thirty you'll probably make it if you do well on the team round. Qualifying as an individual is much harder. It sometimes takes a score of around 40, which means that even star students can fail if they make a few dumb mistakes. My advice would be to have your teammates do well so you don't need to worry about this.

In Massachusetts the state meet is on the same scale as the chapter meet, but with much tougher competition. They give trophies and scholarships to the winners. There's also a big permanent trophy with an engraving that recognizes Bigelow's third place finishes in 2007 and 2010. Winners here advance to the "National" competition in May. The only way to get to the Nationals is with a high individual score. Only four kids from each state get to go. In some states you can get to Nationals with a score in the low 20's. But four is a very small number in a state like Massachusetts. Before Josh Frost moved to Lexington and Idea Math opened, it seemed like you'd need a score of about 38. In the last couple years kids scoring 41 and 43 have found themselves losing out on tiebreakers.

The MathCounts Sprint round is nicely designed to accomplish something very difficult: it's supposed to be a fun and challenging contest that can be taken by kids of different ages and vastly different levels of mathematical preparation. Many questions are designed so that you can solve them via a brute force method even if you don't know the more sophisticated approach that would be the best way to solve them. And the fact that they give you so much less time than most kids would need to do all the problems actually makes it forgiving: if there's some things you don't know you can just skip the questions and do other problems instead. This, however, mostly just means that bright kids from any grade and any background can do pretty well at MathCounts. Winning a competitive chapter or state is another story. For that you need to see the best way to do most problems and to carry out the best method quickly.

If you've been participating in IMLEM all year you'll have a big leg up on students/teams who just start preparing for MathCounts a few weeks before it comes up. MathCounts covers so much material that it's almost impossible to prepare for it by studying math. Unless you've been doing IMLEM. IMLEM and MathCounts cover almost exactly the same set of topics, so what you've been doing to get ready for IMLEM is just what you'd want to do to get ready for MathCounts. In fact, the first thing you should do to get ready for MathCounts is not to read this chapter. First, read the IMLEM Meet #4 sections. You'll need to know them soon anyway, and almost all the topics come up on MathCounts. Then, read the Meet #5 Arithmetic section. The combinatorics and probability topics it covers come up a lot on MathCounts too. Then, read this chapter. Parts of it will be important. Other parts won't. I'd like to tell you which ones to study and which to skip but I can't: the topic on MathCounts vary from year to year.

The best way to try to figure out which parts are important is probably to work though the problems in the *MathCounts Handbook*. I've never done a systematic study of this, but it seems to be a preview of the contest. If some topic comes up multiple times in the *Handbooks's* practice problems, it would be a good idea to learn it on the assumption that there's a good chance it will come up on one or more of the MathCounts meets. Each year's *Handbook* seems to include some problems on topics I haven't seen them cover before, so you can't rely on this book (or any other) to tell you everything you'd like to know.

And the most important thing I can say about getting ready for MathCounts is that it really, really rewards people who practice. To do well at MathCounts you need to know the material AND you need to be very fast. Doing lots of practice problems makes you much faster on the things you know and helps you figure out what you don't know. You should start with the "warmups" and "workouts" in the *MathCounts Handbook*. It takes a long time to work through all of them. Practicing on old Sprint and Target rounds is also really valuable. One unfortunate aspect of the MathCounts program is that they have decided to try to use old problems a funding source. You can't blame them for this – they run the program on a shoestring budget; their few paid employees get low salaries; and even so they ended up losing about $500,000 last year because registration fees only cover about one-third of the costs and they weren't able to attract enough donations to make up the difference. But it does mean that kids at elite MathCounts schools have an advantage: their coaches have dozens of old tests they can use for practice. If you're not at such a school you could buy a decade's worth of old contests from the MathCounts website, but few kids do – they would charge you about $250 for them. You could instead sit around and feel sorry for yourself and complain about how unfair it is. But that won't really do much good. So what I'd recommend instead is to view this as an opportunity. You could buy some books that the kids from the top schools won't bother to buy. (In addition to the AOPS books, Jason Batterson has a book that has a lot of problems and goes over material similar to this book, and Josh Frost has a nice book of MathCounts-style tests he made up.) You can start building your own file with last year's tests, which are free on the MathCounts website. You can find other tests that kids have written online. And you can make up your own practice tests and exchange them with friends. They may not exactly match what's on MathCounts, but the process could end up being even more educational.

Geometry

MathCounts meets have lots of geometry problems on them. Having done IMLEM will be a big advantage because all of the IMLEM topics seem to come up: there are problems on areas and perimeters, Pythagorean Theorem problems, problems about inscribed angles in circles, problems about volumes of pyramids, etc.

Although most MathCounts geometry problems are on things covered on IMLEM, there's enough geometry on MathCounts to make a discussion of the leftover geometry topics the longest section of these notes. The main thing that is on MathCounts but not on IMLEM is analytic geometry – using algebraic equations to describe geometric figures. One other time-saving formula I'd recommend memorizing is the surveyor's area formula. It seems to come up once a year.

2.1 Special Triangles

The last few Mathcounts Handbooks have had several problems involving special triangles. There are two special triangles you should recognize: the 45-45-90 right triangle and the 30-60-90 right triangle.

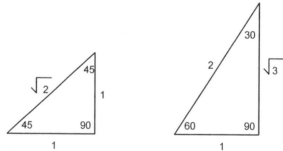

The main thing to know about special triangles is the lengths of the various sides. In a 45-45-90 triangle, the length of the hypotenuse is $\sqrt{2}$ times the length of each leg. For example, if the legs of the triangles are 3 units then the hypotenuse is $3\sqrt{2}$ units. If the length of a leg is $\sqrt{2}$ units, then the length of the hypotenuse is 2 units.

In a 30-60-90 triangle, if the shortest side is x units, then the other leg will be $x\sqrt{3}$ units and the hypotenuse will be $2x$ units.

The simplest problems on special triangles tell you the angles and the length of one side and ask you for another side. More often, however, there's some slight complication. Here are a couple examples in which the complication is that they don't mention the angles.

Question: Square ABCD has perimeter 4. Point E is chosen so that ACE is an equilateral triangle. What is the area of triangle ACE.

Question: Let ABC be a right triangle with hypotenuse AC. Suppose that AB=5, AC=10, and D is the point where the bisector of angle A intersects side BC. What is the length of AD? Give your answer in simplest radical form.

The answer to the first question is $\sqrt{3}/2$. Each side of the square has length 1. AC is a diagonal of the square. It and two sides of the square form a 45-45-90 triangle, so its length is $\sqrt{2}$. The quick way to finish from here is to remember that the area of an equilateral triangle with side length s is $s^2\sqrt{3}/4$. Another equally good but slower way is to rederive the formula for the area of the equilateral triangle by drawing an altitude from point E to side AC. The altitude divides triangle ACE into two 30-60-90 triangles. Each has a short leg of $\sqrt{2}/2$, so the altitude is $\sqrt{3}\sqrt{2}/2 = \sqrt{6}/2$. The area follows by a standard base times height calculation.

The second problem is harder. First, you have to recognize that ABC is a 30-60-90 triangle rather than being told what the angles are. The reason you know that it's a 30-60-90 triangle is that this is the only right triangle where one leg is half the length of the hypotenuse. Angle BAC is the 60 degree angle. The fact that AD bisects angle BAC implies that BAD is a 30 degree angle. So BAD is also a 30-60-90 and you can use your 30-60-90 facts again. The long leg of this triangle is 5 units long, so the hypotenuse, BD is $5\times(2/\sqrt{3}) = 10/\sqrt{3} = 10\sqrt{3}/3$.

If you didn't notice that ABC was a 30-60-90 triangle you could also have done this problem using the Angle Bisector Theorem:

- In triangle ABC suppose that D is a point on BC with AD the bisector of angle A. Then,
$$\frac{BD}{AB} = \frac{CD}{AC}.$$

In this problem the angle bisector theorem tells us that BD/5=CD/10, which gives BD = (1/3) BC = (1/3) $5\sqrt{3}$. (If you didn't recognize that ABC was a 30-60-90 triangle you still could have found BC using the Pythagorean Theorem.) Now applying the Pythagorean Theorem to ABC we have AD^2 = $(5\sqrt{3}/3)^2 + 5^2 = 25(4/3) = 100/3$. I don't know that I've ever seen a MathCounts problem where you needed to know the Angle Bisector Theorem, but it comes up a lot on high school contests so you'll need to learn it at some point.

They also sometimes have contrived word problems that bring in special triangles. For example, a hard question of this type would be:

Question: Handyman Bob is trying to change a light bulb in the ceiling of an eight foot wide hallway. He doesn't have a step-ladder, but does have two regular ladders and figures he could try leaning them against each other. He puts the base of one ladder touching the wall on the left side of the hall and the other ladder touching the wall on the right side of the hallway. When he leans the two ladders together they touch each other exactly at the ceiling, which is nine feet high. The ladder on the right side of the hall makes a 60 degree angle with the floor. What is the length of the longer ladder? Give your answer in feet and inches to the nearest inch.

The first step here is to use the information they've given you about a 30-60-90 triangle. Think about the triangle going from the point where the two ladders meet, down to the floor, and then over to the right wall. It is a 30-60-90 triangle. It's long leg is 9 feet long, so it's short leg is $3\sqrt{3}$ feet long and its hypotenuse (the length of one of the ladders is) $6\sqrt{3}$ feet long. Next you have to figure out which ladder is longer. If you draw a good diagram you will quickly realize that it is the ladder that's $6\sqrt{3}$ feet long. We saw above that the short leg of the triangle is $3\sqrt{3}$ feet or a little more than five feet long. The hallway is only 8 feet wide, so this means that the other ladder is more vertical and shorter. (You can see that it's shorter by using the Pythagorean theorem.) The final step is to figure out what $6\sqrt{3}$ feet is in feet and inches. Hopefully you remember that $\sqrt{3}$ is about 1.732. Multiplying this by 6 we get 10.39 feet or approximately 10 feet and five inches.

Here's one final and again hard problem.

> *Question: In triangle ABC the length of side AB is 12 inches, and the measures of the angles A and B are 45 degrees and 60 degrees, respectively. What is the length of the altitude from vertex C to side AB?*

The way they try to confuse you in this problem is to give you a 45-60-75 triangle. But this is a clue: you don't know anything about 45-60-75 triangles, so it must be that there's something you can do to analyze it using 45-45-90 and/or 30-60-90 triangles. The way to do this is to draw in the altitude from C to AB (meeting AB at D), you'll notice that it divides ABC into two triangles. ACD a 45-45-90 triangle and CDB is a 30-60-90 triangle. The altitude CD is a side of both triangles. Writing x for the length of this altitude we have that AD is also x inches long and DB is $x/\sqrt{3}$ units long. We know that $x + x/\sqrt{3} = 12$, so $x = 12/(1+1/\sqrt{3}) = 12\sqrt{3}/(\sqrt{3}+1)$. The problem asks for the answer in simplest radical form. The way to put this in simplest radical form is to multiply the top and bottom of the fraction by $\sqrt{3}-1$. The answer is $18 - 6\sqrt{3}$.

One way in which they could have made this problem harder would have been to ask for the area of the triangle and not mention the altitude at all. This seems too hard, however, except maybe for one of the last couple problems on a state target round.:

2.2 Analytic Geometry: Lines

2.2.1 Slope

The *slope* of a line is defined by slope = (change in y)/(change in x). For example, a line that goes through the points (3,2) and (4,6) has slope (6-2)/(4-3) = 4.

Slope is a measure of steepness. A very steep upward sloping line has a big positive slope. A flat horizontal line has a slope of zero. A line that slopes gradually downward has a slightly negative slope.

An important property of lines is that the slope is always the same regardless of which two points you use to measure it. You use this fact to answer questions like:

Question: For what value of z do the points (1, 2), (3, 7) and (4, z) line on a line?

Question: For what value of z do the points (1,1), (z,7), and (z+2,5) line on a line?

To do the first problem, you note that the fact that (1,2) and (3,7) lie on the line imply that the slope is 2.5. A line that has slope 2.5 and passes through (3,7) will pass through (4,9.5). The answer is 9.5.

To do the second its good to step back and think about what facts are most useful. The fact that (z,7) and (z+2,5) are both on the line imply that the slope is -1. A line with slope -1 will go through (z,7) and (1,1) when $z = -5$.

2.2.2 Equations for lines

Equations for lines can be put in a few different standard forms. The most common is the slope-intercept form:

- A nonvertical line can be described by the equation of the form $y = mx + b$. The slope of this line is m and the y-intercept is b.

Saying that the "y-intercept is b" means that the line goes through the point $(0,b)$. You can put the equation for a line in this form by finding the slope and y-intercept. For example, suppose you were asked:

Question: The line y=mx+b passes through the points (1, 2) and (3, 7). Find m and b?

One way to do this problem is to first realize from the two points that the slope is 2.5 and then to say that if the line goes through (1,2) and has slope 2.5 it must go through (0,-0.5). Hence the equation is $y = 2.5x - 0.5$.

You can find this equation in a other ways too. One is to first find the slope and then say that the fact that (1,2) is on the line means that 2 = (2.5) 1 + b.

The second standard form for a nonvertical line is given by:

- The equation for the line passing through the point (x_1, y_1) and having slope m is $y - y_1 = m(x - x_1)$.

You can convert between the forms by using standard algebra. For example,

Question: The line y = mx + b passes through the point (2, 5) and has slope 3. Find b.

The answer is -1. You can do it by inspection or write $y - 5 = 3(x - 2)$ and simplify. Sometimes lines are given in other forms like $2x + 3y = 5$. I usually convert these to the $y = mx + b$ form for questions where this makes the problem easier to think about.

The *x-intercept* of a line is the point where it intersects the x-axis, i.e. the point where y=0. You can solve for this using either form of the equation for a line.

> *Question: Find the x-intercept of the line $2x + 3y = 10$.*

The answer is 5. You find it by substituting 0 for y in the equation for the line and solving for *x*.

2.2.3 Distances along lines

Sometimes problems ask you to find a point that is some distance along a line segement. For example,

> *Question: Let A be the point (1, 3) and let B be the point (4, 12). Point E is on segment AB and AE: EB = 1:2. Find E.*

You might think that do this problem you need to write the equation for the line containing AB, then use the distance formula to find AB, then set AE equal to two-thirds of this, then solve for the ordered pair (x, y) that lies on the line at the desired distance from A. You could do the problem this way, but it would be hard and completely unnecessary. The point that is one-third of the way from A to B on the segment AB is simply the point that is one-third of the way there in the x direction and one-third of the way there in the y direction. In this case, it's (2, 6).

2.2.4 Perpendicular lines

A property of lines that comes up pretty frequently is:

* If two lines are perpendicular, then the product of the slopes is -1.

An example of a problem where this would be useful is:

> *Question (Warmup 15-10): What is the intersection point of the line $y = 3x - 2$ and the line perpendicular to it that passes through the point (6,6)?*

The line perpendicular to the given line has slope -1/3. Hence, its equation is $y - 6 = -\frac{1}{3}(x - 6)$ or $y = -\frac{1}{3}x + 8$. The intersection point is the solution to the system

$$y = -\tfrac{1}{3}x + 8$$
$$y = 3x - 2$$

The answer is (3, 7).

2.2.5 Point-to-line distance

The point-to-line distance formula is another must-know formula for math contests. It will come in handy over and over again, not just in MathCounts but also throughout high school. Here's a typical example of a question where it saves you a lot of work.

> *Question: What is the distance between the point (6, 6) and the closest point on the line y = 3x − 2.*

The shortest distance between a point and a line is the length of the segment through the point that is perpendicular to the given line. Hence, one can use formulas for perpendicular lines to find the distance between a point and a line. For example in the problem above you could first say that the slope of the perpendicular is -1/3, then find the intersection of the perpendicular and the original line to find that the closest point is (3, 7), and then use the standard distance formula to find that the distance between (6, 6) and (3, 7) is $\sqrt{3^2 + 1^2} = \sqrt{10}$.

The formula you can memorize to do problems like this more quickly is:

- The distance from the point (x_1, y_1) to the line $ax + by + c = 0$ is $\dfrac{|ax_1 + by_1 + c|}{\sqrt{a^2 + b^2}}$.

To apply the formula in this problem you'd rewrite the equation for the line as $3x − y − 2 = 0$ and then compute the distance to the point (6, 6) as $\dfrac{|3 \cdot 6 − 1 \cdot 6 − 2|}{\sqrt{3^2 + 1^2}} = \dfrac{10}{\sqrt{10}} = \sqrt{10}$. Using the formula is not only a lot quicker but also gives your brain a rest – instead of thinking about the steps and keeping track of where you are you just plug in some numbers.

It won't come up in MathCounts, but another nice aspect of the formula is that it works in higher dimensions:

- The distance from the point (x_1, y_1, z_1) to the plane $ax + by + cz + d = 0$ is $\dfrac{|ax_1 + by_1 + cz_1 + d|}{\sqrt{a^2 + b^2 + c^2}}$.

As I said, this won't come up in MathCounts so you shouldn't worry about this if you don't know what a plane is or find thinking about equations for planes confusing.

2.2.6 Reflections of lines

Reflections aren't as important. They seem to come up in some years. Some facts are:

- If a line is reflected about the *x*-axis (or any horizontal line) or about the *y*-axis (or any vertical line) its slope becomes the negative of what it used to be.

- If (x, y) is reflected about the vertical line $x=a$ it becomes $(2a - x, y)$. A special case is that if (x, y) is reflected about the y- axis it becomes $(-x, y)$.

- If (x, y) is reflected about the horizontal like $y=b$ it becomes $(x, 2b-y)$. A special case is that if (x, y) is reflected about the x-axis it becomes $(x, -y)$.

- If (x, y) is reflected about the line $y=x$ it becomes (y, x).

Here's one problem on reflections.

Question: A square with vertices (1, 1), (5, 1), (1,5) and (5, 5) is reflected about line L. The image is a square with vertices (-7, 1), (-3, 1), (-7, 5), and (-3, 5). What is the equation of line L?

Looking at the square and its image you should notice that the y coordinates are unchanged. This means that the reflection is a reflection about a vertical line of the form $x = a$. A trickier part of the question is that you need to think about which points in the image correspond to which points in the original square – the correspondence is not the order in which I listed the vertices. If you draw the figures you'll realize that the left points on the square with $x = 1$ must be the ones being reflected to the points with $x = -3$ and the right points with $x = 5$ must the ones that are reflected to $x = -7$. It's then obvious that the answer is that the line L is $x = -1$.

I would guess that most reflections will involve lines that are vertical, horizontal, or along the 45 degree line. If you need to do a reflection along some other line, however, you can always take the brute-force approach: a reflection is a movement along the line perpendicular to the given line such that the average of the original point and the image is equal to the intersection of the line-of-reflection and the perpendicular. Don't try doing this, however, unless it's a team round question or a target round question in which you've already gotten the other target problem right.

2.3 Analytic Geometry: Circles

There were more question on circles in the 2007-2008 Handbook than in the next couple after that, but it's still probably good to know at least a little about equations for circles.

2.3.1 Equations for circles

The main fact here is:

- The equation for a circle with center (x_0, y_0) and radius r is $(x - x_0)^2 + (y - y_0)^2 = r^2$.

MathCounts sometimes asks you about this by giving you an equation like this and not telling you it's a problem about circles. For example, they could ask:

Question: Find the area of the region satisfying $(x - 3)^2 + (y + 2)^2 \leq 9$, $x \leq 3$, and $y \leq -2$.

If you're familiar with the equation for a circle this is an easy problem. It's just asking you for the area of the lower left quarter of a circle with radius 3. The answer is 9/4 π. If you don't recognize the equation for a circle, you'd just have to skip this one, which is why I think it's good to at least know the basics of this section.

2.3.2 Tangents

A *tangent* to a circle is a line that is touches the circle at one point. This turns out to be the line that is perpendicular to the line connecting the center of the circle to the point on the edge of the circle. For example, the line tangent to the circle $x^2 + y^2 = 2$ at the point $(1,1)$ has slope -1 because the line connecting $(0,0)$ and $(1,1)$ has slope 1. To think about this graphically, note that the tangent line is just the line connecting $(0, 2)$ and $(2, 0)$. Another example is that the circle $x^2 + y^2 = 49$ is tangent to the line $y = 7$ at the point $(0,7)$.

An easy question about tangents would be:

Question: What is the area of the circle that is centered at the origin and tangent to the line x=5.

To do this one, you just note that the circle has radius 5 and hence has area 25π. An example of a medium question is:

Question: A circle with center (4, 0) is tangent to the line y = x. What is the radius of the circle? Express your answer in simplest radical form.

One way to do this is to draw a quick diagram containing the circle, the line $y = x$, and a line connecting the center of the circle to the point of tangency. Noting that the two lines are perpendicular you should realize that the triangle with vertices at $(0, 0)$, $(0, 4)$, and the point of tangency is a 45-45-90 triangle. The answer is therefore $4/\sqrt{2} = 2\sqrt{2}$. Another way to do this problem would have been with the point-to-line distance formula. The distance from the center of a circle to any tangent line is the radius of the circle.

A harder question is:

Question: Circles with centers at (0, 4) and (3, 0) are both tangent to the line y=x. What is the distance between the closest points of the two circles? Express your answer as a common fraction in simplest radical form.

The first step on this one is to draw a good diagram. This should make it obvious that the answer is not zero: one circle lives below the line, the other lives above the line, and they never touch. The second harder step is to realize that the closest points on two circles are the points where the circles intersect the lines joining their centers. Once you do this, the problem is just a medium-difficult problem. The first circle has radius $4/\sqrt{2}$ as we saw in the last problem. The second has radius $3/\sqrt{2}$. The distance between the centers of the two circles is $\sqrt{3^2 + 4^2} = 5$. The segment connecting the two centers is divided into three parts by the circles. The left part inside the big circle has length $4/\sqrt{2}$. The right part inside the small circle has length $3/\sqrt{2}$.

The remaining part in the middle has length $5 - 7/\sqrt{2} = \dfrac{10 - 7\sqrt{2}}{2}$. This is the answer.

2.4 Areas of Polygons on a Grid

Polygons on a grid are another of my favorite MathCounts topics. There are a couple of neat techniques that make hard-sounding problems remarkably easy. A typical example is

Question: If O is at (0,0), A is (4,7) and B is (5,4), find the area of triangle OAB.

There are a few ways to do problems like this. One would be to find the lengths of each of the three sides using the Pythagorean theorem and then to use Heron's formula to find the area. It turns out, however, that this is usally a very bad idea. In this problem, for example, the sides have lengths like $\sqrt{41}$, which makes the semiperimeter kind of complicated and the full formula a really big mess. There are three other ways to do problems like this that are much easier.

2.4.1 Subtraction method

To understand this method, get some graph paper and label the points O, A, and B. Then, draw two more points on your grid: put point C at (5,0) and point D at (4,0). We then have

Area(OAB) = Area(OABC) – Area(OBC) = Area(OAD) + Area(DABC) – Area(OBC).

Each of the three areas on the right side is easy to compute. OAD is a right triangle with area ½ (4 × 7). DABC is a trapezoid with area ½ (7 + 4) × 1. OBC is a right triangle with area ½ (5 × 4). Hence the answer is 14 + 5.5 – 10 = 9.5.

2.4.2 Surveyor's area formula

Surveyors are used to drawing plots of land by finding the latitude and longitude of each of the vertices marking the boundary of the property and then connecting them. They are also often asked to compute the size of plots in acres (or some other unit.) As a result, they long ago noticed a simple formula for computing areas. In the case of triangles the formula is:

Theorem: The area of a triangle with vertices at (0,0), (x_1, y_1), and (x_2, y_2) is
$A = \frac{1}{2} |x_1 y_2 - x_2 y_1|$

In our problem the formula gives A = ½ |(5 ×7) − (4 × 4)| = (35 − 16)/2 = 9.5, which of course is the same answer as before. You could prove to yourself that the formula works for any triangle by drawing a diagram and using the subtraction method.

The surveyor's formula also has a generalization that lets you just as easily compute the area of any simple polygon:

Theorem: Suppose that the vertices of a simple n-gon as you go around it counterclockwise are in order (x_1, y_1), (x_2, y_2), ..., and (x_n, y_n). The, the area is

$$A = \frac{1}{2}\left((x_1 y_2 - x_2 y_1) + (x_2 y_3 - x_3 y_2) + ... + (x_{n-1} y_n - x_n y_{n-1}) + (x_n y_1 - x_1 y_n)\right)$$

This may look like a complicated formula, but there's actually a pattern that makes it easy to remember. Here's how it goes. First, you write down all the points in a list (in counterclockwise order), putting the first one both at the top and at the bottom. Then, you multiply the diagonal terms in each pair and compute the difference between the downward product and the upward product. Finally, you add up all the numbers in the right column and divide by two.

(x_1, y_1)	$\rightarrow$	$x_1 y_2 - x_2 y_1$
(x_2, y_2)	$\rightarrow$	$x_2 y_3 - x_3 y_2$
(x_3, y_3)	$\rightarrow$	$x_3 y_4 - x_4 y_3$
...	...	
(x_n, y_n)	$\rightarrow$	$x_n y_1 - x_1 y_2$
(x_1, y_1)		_____

If you draw lines on the paper connecting the things you multiply you'll notice that the pattern looks like a pair of shoelaces. I've heard people call this the "shoelace" area formula. Here's an example:

Question: Find the area of the quadrilateral with vertices (0, 0), (1, 4), (5, 5), and (4, 1).

To use the surveyor's formula you first need to put the points in counterclockwise order. This is (0, 0), (4, 1), (5, 5), and then (1, 4). The computation is then

(0, 0)	$\rightarrow$	0 − 0 = 0
(4, 1)	$\rightarrow$	20 − 5 = 15
(5, 5)	$\rightarrow$	20 − 5 = 15
(1, 4)	$\rightarrow$	0 − 0 = 0
(0, 0)		_____
		30 ÷ 2 = 15.

There are several ways to verify this answer if you're skeptical. One is to divide the rhombus into two triangles and use the surveyor's formula on each. Another is to find the area using the fact that the diagonals of a rhombus are perpendicular and divide it into four right triangles.

The formula doesn't just work on quadrilaterals. It works on pentagons, hexagons, etc. and saves even more time there. One additional tip when using the surveyor's area formula is that you can make the calculations easier by translating the figure so that that one of the points is at (0, 0). For example if I was asked,

Question: Find the area of the quadrilateral with vertices (3, 7), (4, 11), (8, 12), and (7, 8).

the first thing I'd do would be to say that the answer remains the same if I subtract 3 from the x-coordinate of each point and 7 from the y-coordinate of each point. This means the area is the same as the area of the quadrilateral with vertices (0, 0), (1, 4), (5, 5), and (4, 1). As above the answer is 15. Shifting one of the points to (0, 0) helps for two reasons: the numbers involved in the multiplications and subtractions are smaller; and two of the "shoelace" products always turn out to be zero.

2.4.3 Pick's formula

If you're given a picture of a polygon drawn on graph paper and all of the vertices of the polygon are at grid points (i.e. their coordinates are all integers), then Pick's formula provides a simple way to find the area. If the figure is relatively small it can be a very quick way to get the solution the theorem is:

Theorem: Suppose that the vertices of a simple polygon are all integer points on a grid. Then the area of the polygon is

$$A = I + ½ B - 1,$$

where I is the number of grid points in the interior of the figure, B is the number of grid points on the boundary of the polygon and 1 is the number one.

Going back to the problem at the start of this section about finding the area of a triangle with vertices at (0,0), (4,7), and (5,4), if you draw an accurate picture for my triangle problem you should see that there are nine grid points inside the triangle – (1,1), (2,2), (2,3), (3,3), (3,4), (3,5), (4,4), (4,5), and (4,6) – and that the three vertices are the only grid points on the boundary. Hence the area is $9 + ½ \cdot 3 - 1 = 9.5$.

Another problem in which it would have been very easy to get the answer this way is in the octagon I showed in section 2.2 of the part on IMLEM Meet #2: there are four grid points inside the shaded region and eight on the boundary so the area is $4 + ½ \cdot 8 - 1 = 7$.

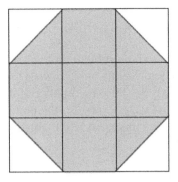

You could even have used this method to find the area of the big fish-shaped polygon in the first illustration of that chapter. By the time a figure gets this big though, I think of Pick's theorem as not being a great approach. It may still be quicker than doing the problem by subtracting the various triangles from the square, but it's easy to get lost while counting the points and to end up with an answer that's off by one.

2.4.4 Counting grid points via Pick's formula

Counting grid points in a figure is another classic MathCounts topic. It's attractive to them because any kid can solve the problems by just counting all the points, and some kids will get the problems more quickly by finding a pattern. There's a third even better way to do some of these problems using Pick's theorem. Questions where you could do this are things like

Question: How many points with integer coordinates are strictly inside the rectangle with vertices (0,0), (9,0), (0, 4), and (9, 4)?

Question: How many points with integer coefficients lie inside the triangle with vertices (0,0), (10, 1) and (5,2)?

Question: How many ordered pairs of integers (x, y) satisfy $5|x| + 2|y| < 20$.

The first of these you can obviously do just by counting the points. They're obviously a 3×8 grid, with x-values ranging from 1 to 8 and y-values ranging from 1 to 3 so the answer is 24.

For problems like the second one, however, counting the interior points is harder. Here, a combination of Pick's Theorem and the surveyor's area formula provides an easier approach. Pick's Theorem says that $\frac{1}{2}B + I - 1 = A$. Or equivalently that the number of grid points inside the triangle is $I = A - \frac{1}{2}B + 1$. The surveyor's area formula gives that the area of the triangle is $\frac{1}{2}(10 \times 2 - 1 \times 5) = 7.5$. B is 3 because the sides of the triangle contain no grid points other than the vertices (you can see this by thinking about whether the change in y and the change in x on each side have any common factors.) Hence, $I = 7.5 - 1.5 + 1 = 7$. This may not seem so much better than counting the points in this problem, but imagine how hard it would be to count the points directly if the vertices of the triangle were (0, 0), (17, 3), and (8, 25)! Another good thing about knowing methods like this is that doing a problem by a second completely different method is a good way to check your work (which you only really do on the Target and Team Rounds.)

The third problem might seem like an algebra problem, but it's also a geometry problem. The inequality in the question is defining the rhombus with vertices (4, 0), (0, 10), (-4, 0), and (0, -10). The area of this rhombus is $\frac{1}{2} \times 20 \times 8 = 80$. The number of boundary points is 8: the four vertices plus the four points of the form (± 2, ± 5). Hence, the number of grid points in the interior is $80 - \frac{1}{2} 8 + 1 = 77$. If the problem had said ≤ 20 then you'd need to add the 8 boundary points too.

2.5 The Triangle Inequality

One other fact that comes up occasionally is the triangle inequality:

- *If a, b, and c are the lengths of the sides in a triangle then $a + b > c$.*

This should be intuitive. Think of attaching segments of length a and b to hinges at the opposite ends of a segment of length c. If $a + b < c$, then these two segments won't reach each other and can't form a triangle.

The simplest Mathcounts problems ask about this almost directly.

Question: Two sides of a triangle have lengths 3 and 5. What is the largest possible integer value for the third side?

The only trick to this one is that the answer is 7 and not 8. They also like using the triangle inequality in connection with some counting problem. For example, they might ask:

Question: How many non-congruent triangles, each with a perimeter of 13 units, can be constructed with all integral side lengths.

Like many Mathcounts problems, this isn't one where you're expected to know some formula. Instead, you're just supposed to count all the possibilities by keeping a logically ordered list. I think the best way to proceed in this one is by looking in order at all possible values for the length of the shortest side:

If the shortest side has length 1, then the other two sides must be 6 and 6. Any other combinations that add to 12, e.g. 7+5, are not possible because the shortest two sides are too short.

If the shortest side is 2, then there is again one possibility for the remaining sides: 2-5-6.
3 gives two possibilities: 3-5-5 and 3-4-6.
4 gives one more: 4-4-5.

The answer is 5.

2.6 A Few Random Facts

The MathCounts Handbook includes an extremely short list of formulas. The list is so short that I thought I should mention one formula on it that I've only mentioned in passing.

- *The area of a rhombus is $A = \frac{1}{2} d_1 d_2$, where d_1 and d_2 are the lengths of the diagonals.*

The reason why this is true is that the diagonals of a rhombus are perpendicular. A second random fact is an area for a triangle:

- *The area of triangle ABC is $A = \frac{1}{2} ab \sin(C)$.*

If you don't know any trigonometry, I wouldn't try to learn it. But this formula comes in handy for answering questions like.

Question: Find the area of parallelogram ABCD the length of AB is 10 units, the length of BC is 8 units and angle ABC is 60 degrees.

The way to solve this problem without using the above formula is to draw the altitude from C to point E on side AB, and then to use the fact that EBC is a 30-60-90 triangle to find that the height of the parallelogram is $4\sqrt{3}$, which makes the area $40\sqrt{3}$.

Number Theory

There isn't a whole of number theory on MathCounts. They seem to like geometry, arithmetic, combinatorics, and simple algebra much more. Many of the number theory problems they do have are things you should know well if you've done IMLEM, e.g. they ask about how many factors a number has. The one number theory topic I thought I'd say more about is modular arithmetic. If you're busy though, don't read this section. It's much less likely to be useful than the others.

3.1 More Modular Arithmetic: The Fermat-Euler Theorem

One example of a modular arithmetic question that you should already know how to do is:

Question: What is the remainder when 2011^{2010} is divided by 5?

The problem will really impress a parent who doesn't know modular arithmetic, but is not at all hard if you remember basic facts from the IMLEM Meet #4 number theory section: $2011 = 1$ (mod 5), so you just need to compute 1^{2010} (mod 5), which is 1. If you were instead asked:

Question: What is the remainder when 2012^{2011} is divided by 5?

it would be useful to remember a theorem I discussed toward the end of that section called Fermat's Little Theorem:

- If p is prime and a is not a multiple of p then $a^{p-1} = 1$ (mod p).

This theorem tells us that $2^4 = 1$ (mod 5). Hence, $2012^{2011} = 2^{2011} = (2^4)^{502} 2^3 = 3$ (mod 5). You don't necessarily need to remember the theorem because you could have done the problem by multiplying out powers of 2 mod 5 until you see the pattern (which is that it repeats after four terms), but remembering the theorem helps.

Problems about units digits are more common on MathCounts. In 2013 they might ask:

Question: What is the units digit when 2013^{2012} is written out as a decimal number?

Problems asking about units digits are asking you to simplify things mod 10. You can't use the theorem above, because it's about simplifying numbers in mod p arithmetic when p is prime. Ten is not prime.

One way to find the units digit of 2013^{2012} is just to recognize that it's same as the units digit of 3^{2012} and then start multiplying 3's together until you see the pattern: $3^1 = 3$ (mod 10), $3^2 = 9$ (mod 10), $3^3 = 27 = 7$ (mod 10), $3^4 = 3 \times 3^3 = 3 \times 7 = 1$ (mod 10). From here, you see that the pattern repeats after every four terms and know that the units digit of 3^{2012} is a one.

An easier way to do this part of the problem is to use the Fermat-Euler theorem:

- If GCF(a,m)=1 then $a^{\varphi(m)} = 1 \pmod{p}$.

I'll tell you more about the function $\varphi(m)$ below. For now, just remember that $\varphi(10)=4$ and $\varphi(100)=40$. This is enough to get the two most important corollaries for Mathcounts:

- If a is odd and not a multiple of 5, then $a^4 = 1 \pmod{10}$.

- If a is odd and not a multiple of 5, then $a^{40} = 1 \pmod{100}$.

The first tells you that the fact that $3^{2012} = 1 \pmod{10}$ is not a coincidence. You'd also get one if you simplified 7^{2012} or 9^{2012} mod 10.

One thing to be careful about here is that this theorem does require that a be odd. To evaluate 2^{2012} you'll need to use the looking-for-a-pattern method.

Remembering that $\varphi(10)=4$ and $\varphi(100)=40$ makes some problems remarkably easy. For example,

Question: What is the tens digit of 37^{41}?

Here, it is immediate from $\varphi(100)=40$ that $37^{41} = 37^1 = 37 \pmod{100}$. The answer is 3.

3.1.1 The phi function

One thing that can make advanced mathematics daunting is that mathematicians start to use Greek letters for everything instead of regular letters. Most kids have seen the symbol π sufficiently frequently so that it doesn't bother them any more. The other ones won't be confusing either once you're used to them, but I know that for now it is annoying.

The greek-letter-named function that appeared in the equation above is the phi function, written $\varphi(m)$. The definition is:

- $\varphi(m)$ is the number of positive integers less than m that have no common factors with m.

This function is easy to evaluate for prime numbers: if p is prime, then $\varphi(p) = p - 1$. For other numbers there's a formula that makes it also not very hard to evaluate:

- If the prime factorization of n is $n = 2^a \cdot 3^b \cdot 5^c \cdot 7^d \cdot 11^e \cdot 13^f \cdot \ldots$, then
$$\varphi(n) = (2-1)2^{a-1} \cdot (3-1)3^{b-1} \cdot (5-1)5^{c-1} \cdot (7-1)7^{d-1} \cdot (11-1)11^{e-1} \cdot (13-1)13^{f-1} \cdot \ldots$$

For example,

$\varphi(35) = \varphi(5^1 7^1) = (5\text{-}1)5^0 \cdot (7\text{-}1)7^0 = 24.$
$\varphi(36) = \varphi(2^2 3^2) = (2\text{-}1)2^1 \cdot (3\text{-}1)3^1 = 12.$
$\varphi(63) = \varphi(3^2 7^1) = (3\text{-}1)3^1 \cdot (7\text{-}1)7^0 = 36.$

This is a very handy formula to know if you're asked something like:

Question: How many positive integers x less than or equal to 36 have GCF(x,36) > 1.

The answer is 36 - $\varphi(36)$ = 24.

I haven't noticed them asking many questions like this in MathCounts. I included this section mostly because it seemed inappropriate not to after I'd given you the Fermat-Euler theorem.

3.2 Repeating Decimals

The last few Mathcounts Handbooks don't have much on repeating decimals, but there were some problems on them in earlier years. This section is kind of long so you should definitely not make reading it a priority, but if you have time there's some chance it could be useful.

Back in the IMLEM Meet #2 section I discussed a theorem that is useful for answering questions like

Question: What is the 34th digit after the decimal point in the decimal expansion of 1/17.

The theorem was

- *Theorem: If p is a prime number (other than 2 or 5) and a is not a multiple of p, then a/p is a pure repeating decimal and the length of the repeat is a factor of p – 1.*

In the question above the theorem tells you that the repeat length is a factor of 16. This implies that the 34st digit in the decimal expansion of 1/17 is the same as the 18th and the 2nd. The decimal expansion of 1/17 starts 0.05882… so the answer is 5.

3.2.1 More on prime denominators

In many problems knowing the above theorem is enough. Sometimes though, you need to know more. For example, suppose you were asked about the 19th digit in the decimal expansion of 1/13. The theorem says that 1/13 is a pure repeating decimal and the repeat length is a factor of 12. We don't know, however, whether the repeat is actually 12 (in which case all we can say is that the 19th digit is the same as the 7th) or whether the repeat is a proper factor of 12 (which would make the answer easier.)

A theorem that can help in some problems like this is:

- *Theorem: If p is prime and 10^k is one more than a multiple of p, (p is not 2 or 5 and a is not a multiple of p), then a/p is a pure repeating decimal and the length of the repeat is a factor of k.*

The reason why this theorem works is because of the 999 method for finding repeating decimals. If 10^k is one more than a multiple of p, then p is a factor of the number 999...99 with k 9's in it.

To apply the theorem to find the repeat length use modular arithmetic. In modular arithmetic the way the way to write that 10^k is one more than a multiple of p is to write $10^k \equiv 1 \pmod{p}$. The way you would find out which k this is true for in the mod 13 case is

$10^1 \equiv 10 \pmod{13}$.
$10^2 = 10 \times 10 \equiv 9 \pmod{13}$.
$10^3 = 10 \times 10^2 \equiv 10 \times 9 \pmod{13} \equiv -1 \pmod{13}$

At this point, you should immediately realize that

$10^6 = 10^3 \times 10^3 \equiv -1 \times -1 \pmod{13} \equiv 1 \pmod{13}$

Hence, 13 does have a repeat length of 6 and the 19th digit of (1/13) is the same as the 1st, which is zero.

3.2.2 Denominators with 2's and 5's

The discussion above was all about repeat lengths for fractions with prime denominators. This may make you wonder what happens when the denominator is not prime. This is sufficiently complicated so that it may not be worth trying to study it, but here are some facts:

First, consider denominators that have some 2's and 5's in them.

- If a reduced fraction is of the form $a/(2p)$, $a/(5p)$, or $a/(10p)$ then the decimal expansion will have one patternless digit after the decimal point and then repeat with the same period as p.

- If a reduced fraction is of the form $a/(4p)$, $a/(25p)$, $a/(20p)$, $a/(50p)$, or $a/(100p)$, then the decimal expansion will have two patternless digits after the decimal point and then repeat with the same period as p.

The reason why these formulas work is you can multiply the fraction by 10 or 100 and get something that's a pure repeating decimal. For example, so solve

Question: What is the 40th digit after the decimal point in the decimal expansion of 1/28?

The first step is to say that $100\,(1/28) = 25/7 = 3\frac{4}{7}$. The decimal expansion of 4/7 is $0.\overline{571428}$, so the expansion of 1/28 is $0.03\overline{571428}$. The 40th digit is the same as the 4th, which is 7.

3.2.3 Why does this work? Repeat lengths and Fermat's Little Theorem

If you've been paying close attention you may have noticed that the two theorems I gave about prime repeat lengths were related.

- *Theorem: If p is a prime number (other than 2 or 5) and a is not a multiple of p, then a/p is a pure repeating decimal and the length of the repeat is a factor of p – 1.*

- *Theorem: If p is prime and 10^k is one more than a multiple of p, (p is not 2 or 5 and a is not a multiple of p), then a/p is a pure repeating decimal and the length of the repeat is a factor of k.*

The relation is that the first is a Corollary of the second. Again, the key is Fermat's Little Theorem:

- If p is prime and a is not a multiple of p then $a^{p-1} = 1$ (mod p).

When p is prime (and not 2 or 5), 10 is not a multiple of p, so Fermat's Little Theorem tells us that $10^{p-1} = 1$ (mod p).

3.2.3 Repeat lengths for most denominators

If the denominator of a fraction doesn't have any 2's and 5's in it, then you can use the Fermat-Euler theorem to find the repeat length. For example:

Question: What is the 100th digit after the decimal point in the decimal expansion of 1/21.

To do this problem, note that the Fermat-Euler theorem says $10^{12} = 1$ (mod 21). Using our theorem on powers of 10, the repeat length is a factor of 12. Hence, the 100th digit is the same as the 4th. Doing some division we find $1/21 = 0.0476...$ The answer is 6.

If you find the phi function hard to remember you could instead just try to remember the simpler case of what happens when you have a product of two primes:

- If a reduced fraction is of the form $a/(pq)$, with p and q both prime and different from 2 and 5, then the decimal expansion will have a repeat length that is the least common multiple of the repeat length of (1/p) and the repeat length of (1/q).

You could also just not bother with anything after the result that the repeat length of a/m is a factor of k if $10^k = 1$ (mod m). In a problem like the one above you can always just start from this and compute

$10^2 = 10 \times 10 \equiv -5 \pmod{21}$.
$10^4 = 10^2 \times 10^2 \equiv -5 \times -5 \pmod{21} \equiv 4 \pmod{21}$
$10^6 = 10^4 \times 10^2 \equiv 4 \times -5 \pmod{21} \equiv 1 \pmod{21}$

This doesn't take so long and it lets you realize that the repeat length is actually 6 (whereas the theorem just tells you it's some factor of 12).

Arithmetic

Combinatorics is one of the main topics they emphasize in MathCounts. Before reading this section you should read the Meet #5 section on combinatorics at least once. The problems I discuss here are more complicated. If you don't know the basics well, they'll confuse you and make you forget what you used to know.

Do try to get to this section though. The topics discussed here seem to come up all the time on state-level MathCounts tests. Many problems that are really hard if you haven't read this chapter are really easy if you have. I'll do one other quick topic before I start the combinatorics.

4.1 Series and Products

Mathcounts meets usually have problems that ask you to add up or multiply sets of numbers. The majority of these involve arithmetic series. Geometric series also come up pretty often. Most of what you need to do to solve the problem is just to recognize whether the series they're giving you is an arithmetic or geometric series. Two examples are:

Question: What is the sum of all positive two digit integers that have a units digit of 3?

Question: A ball is dropped from a height of 90 feet. Each time the ball hits the ground it bounces back up ⅔ of the height from which it just fell. How many total feet does the ball travel by the time it hits the ground the third time?

The first of these is an arithmetic series problem and the second is a geometric series problem. The answers are 477 and 290. If you don't remember how to do these right away (or if you skipped the advanced topics when you first read it) you should reread the Meet #3 Arithmetic section to review.

There are also a couple tricks that seem to come up pretty often that I didn't discuss in the Meet #3 section.

4.1.1 Telescoping series and products

A "telescoping" series is one in which lots of things cancel if you regroup the terms in some clever way. Some examples are:

$$1 - 2 + 3 - 4 + 5 - 6 + 7 - 8 + 9 - 10$$

$$\frac{1}{2} \cdot \frac{2}{3} \cdot \frac{3}{4} \cdot \frac{4}{5} \cdot \frac{5}{6} \cdot \frac{6}{7} \cdot \frac{7}{8}$$

The easy way to do the first one is group the terms together in pairs: $(1 - 2) + (3 - 4) + \ldots$ This makes it obvious that there are five terms each of which is equal to -1. The sum is -5. The thing to do with the product is to recognize that the denominator of each fraction (except the last one) cancels with the numerator of the next fraction. All that's left is 1/8.

The canceling in the above problems is so obvious that it would be hard to miss. The people who make up Mathcounts tests typically try to do something to make it more likely that you'll miss the trick. One thing they do sometimes is to regroup the terms.

Question: Compute (2 + 4 + 6 + ... + 100) – (1 + 3 + 5 + ... 99).

You could do this problem by adding up both arithmetic series and then subtracting. The faster way is to rewrite the problem as $(2 – 1) + (4 – 3) + ... + (100 – 99)$. This makes it obvious that the answer is 50.

Another trick that makes for much harder problems (if you haven't seen the trick) is to add up some of the terms to obscure that a series telescopes. Here's a classic example.

Question: Compute $\dfrac{2}{1 \cdot 3} + \dfrac{2}{3 \cdot 5} + \dfrac{2}{5 \cdot 7} + \dfrac{2}{7 \cdot 9} + \dfrac{2}{9 \cdot 11}$

If you don't know the trick, you could approach this problem with a brute force method. Start adding terms together really fast. You could also use a hybrid of this method and the "look for a pattern" method. Add the first three terms together. Notice that the first term is 2/3, the sum of the first two is 4/5, and the sum of the first three is 6/7. From here you could guess that the answer might be 10/11. If you're short on time you might want to put down this guess instead of continuing to add.

The trick for this problem is very cute. Notice that $\dfrac{2}{k \cdot (k + 2)} = \dfrac{1}{k} - \dfrac{1}{k + 2}$. Hence the series can be rewritten as a series that's longer, but which telescopes:

$$\frac{2}{1 \cdot 3} + \frac{2}{3 \cdot 5} + \frac{2}{5 \cdot 7} + \frac{2}{7 \cdot 9} + \frac{2}{9 \cdot 11} = \left(\frac{1}{1} - \frac{1}{3}\right) + \left(\frac{1}{3} - \frac{1}{5}\right) + \left(\frac{1}{5} - \frac{1}{7}\right) + \left(\frac{1}{7} - \frac{1}{9}\right) + \left(\frac{1}{9} - \frac{1}{11}\right) = 1 - \frac{1}{11} = \frac{10}{11}$$

It's sufficiently clever so that it's unlikely that you'd come up with the trick on your own. Now that you've seen it though, you may get the problem right if they use a similar trick again.

4.1.2 Adding numbers in columns

The second trick is sort of an anti-trick. The trick is to not look for a trick and simply to add the numbers up using the brute force method you learned in first grade. Here's an example.

Question: Twenty four different four digit numbers can be formed by rearranging the digits of the number 4321. What is the sum of these twenty-four numbers.

Many students will be paralyzed when they are first confronted with this problem. They'll think 'I can't add twenty four four-digit numbers together... So there must be some trick ... Is there some formula for this I'm supposed to remember … Is this some kind of series …' and never get around to starting on the problem.

The trick to this problem is just to write down the problem as a really hard first-grade problem and think about what it looks like.

```
   4321
   4312
   4231
   4213
   4132
   4123
   3421
    ⋮
    ⋮
+ 1234
```

What is in the ones column of this addition problem? There will be twenty four numbers. By the inherent symmetry of the problem we know there will be an equal number of 1's, 2's, 3's and 4's. (This is even more obvious if you think about the first column.) So the sum of the digits in the one's column is $6(1 + 2 + 3 + 4) = 60$. Hence, you can write down a zero at the bottom of the one's column and carry the 6.

Now, what's in the tens column? Again, it's six 1's, six 2's, six 3's and six 4's. Hence, adding up all the numbers plus the 6 that was carried from the ones column gives 66. So you write a 6 in the tens column of the answer and carry a six to the next column. From here, the pattern should be obvious. The answer is 66660.

4.2 Review of Standard Combinations Problems

The standard "combinations" problem asks you how many ways there are to choose a group of size k out of a set of size N, e.g.

Question: A math team has 14 students. In how many ways can the coach choose 10 of the students to be official contestants?

The answer is $_{14}C_{10} = \dfrac{14!}{10!4!}$. Other problems ask about dividing a set into two groups, e.g.

Question: In how many different ways can a class of fourteen students be divided into a group of 10 and a group of 4?

Note that the answer to this question is also $_{14}C_{10}$. Once you choose 10 students to be in the first group you have no choice about whom to put in the second group.

The one thing you have to be careful with in these problems is what exactly the question says if two or more of the groups are the same size. Consider for example:

Question: The Bigelow MathCounts team has 8 students. In how many ways can they choose 4 students to be official competitors and 4 students to be alternates?

Question: The Bigelow MathCounts team wants to practice doing team rounds. In how many different ways can the 8 students be divided into two groups of 4?

The answer to the first question is $_8C_4$. If you start by picking four people to be regulars, any group of four you pick is a different plan for the meet.

The answer to the second is not $_8C_4$. It's $_8C_4 / 2$. The groups don't have names in the second problem, so there's an extra level of double-counting you need to think about. In the first problem having A, B, C and D as regulars and E, F, G and H as alternates is different from having E, F, G and H as regulars and A, B, C and D as alternates. In the second problem, having A, B, C and D in the first group and E, F, G and H in the second is **not** different from having E, F, G and H in the first group and A, B, C and D in the second. Hence, if you think of counting plans for dividing the students into groups by picking four students to be in the first group, you'd be counting each division twice.

Note that this complication only comes up if the two groups are exactly the same size. If one group has ten students and one has four, then there's never any confusion.

4.3 Combinations Problems with Three or More Groups

It's easy to come up with combinations problems that have more than two groups. For example,

Question: A class with eight students is putting on a play. Five students will have acting parts, two students will be on the production crew, and one student will be the narrator. In how many different ways can each of the students be assigned to one of these three jobs (actor/production crew/narrator)?

The formula for dividing students into m named groups is very similar to the regular combination formula:

- The number of ways to choose divide a set of N elements into *named* groups of size k_1, k_2, …, and k_m is $\dfrac{N!}{k_1!k_2!...k_m!}$.

The answer to the above problem is that there are $\dfrac{8!}{5!2!1!} = \dfrac{8\times7\times6}{2} = 118$ ways to assign jobs to each student.

One way to think about why this is true is to think about sequential choices. First, you can pick any of the eight students to be the narrator. After, this you can pick any of seven students to be the production crew member #1 and any of the remaining six to be production crew member #2. The five remaining students must then all be actors. We've double counted the assignments, however, because having A as production crew member #1 and B as production crew member #2 is the same as having B as production crew member #1 and A as production crew member #2.

A more systematic way to think about this is to think about random orderings and double counting like we did when first discussing combinatorics in the IMLEM Meet #5 section. We can generate all divisions by picking a random ordering for the students and then assigning the first five to be actors, the next two to be the production crew and the last student to be the narrator. There are 8! different orders we can choose. They do not, however, all result in different assignments of roles. How many different orders correspond to each assignment? Any reordering of the first five people will leave the assignment unchanged, so we know we've counted each allocation 5! times. Also, any reordering of the next two students leaves the allocation unchanged, which gives us a further factor of 2!. Hence, the answer is $\frac{8!}{5!2!}$.

If the groups don't have names you sometimes have to take into account yet another aspect of double counting to get the answer. For example, consider

Question: In how many ways can a group of eight students be divided into a group of four and two groups of two?

The answer to this is not $\frac{8!}{4!2!2!}$ for the same reason as in the problem where I talked about dividing the Bigelow math team into two groups of four. It's $\frac{1}{2}\frac{8!}{4!2!2!}$ because putting x and y in group 2 and z and w in group 3 is the same as putting z and w in group 2 and x and y in group 3. In general, if there are k unnamed groups of the same size you need to divide the earlier formula by $k!$.

4.4 Using Combinations to Count Orderings

One place in which the multi-group combination formula comes in handy is in counting possible orderings. For example,

Question: How many different 6 digit numbers can be obtained by reordering the digits of the number 411212?

Question: How many different four-letter "words" (whether they mean anything or not) can be formed by reordering the letters ANNA.

Question: A class has 20 students. The teacher is required to give out exactly 8 A's, 8 B's, and 4 C's. She reports the grades by writing them on a grade sheet that has the students' names in alphabetical order. How many 20 letter "words" could be formed by the letters on the grade sheet?

These may not seem like combinations problems, but they are. The way to think about the first problem is to recognize that when you're making up an ordering you need to pick which three of the six places you will put a 1 in, which two you will put a 2 in, and which one you will put a four in. Hence, its like you were taking six students (ones-place, tens-place, … , hundred-thousands-place) and assigning them to one of three jobs: the job of displaying a number 1; the job of displaying a number 2; and the job of displaying a number 4. Note that three place-values must be given job 1, two must be given job 2, and one must be given job 4. The answer is

$$\frac{6!}{3!2!1!} = \frac{6 \times 5 \times 4}{2} = 60.$$

The second problem is easier. You just need to pick two of the four places to have the letter A in them. Then answer is $_4C_2 = 6$.

The third problem is another classic three-group combination problem: the teacher is assigning each student the job of being an A-student, a B-student, or a C-student. The answer is $\dfrac{20!}{8!8!4!}$.

4.5 Counting Ordered Ways to Make Numbers Add Up

MathCounts seems to like asking about one other famous counting problem. It's a very cute problem, so I definitely wanted to include it. An example is

Question: How many ordered triples (a, b, c) of positive integers satisfy a + b + c = 10.

A famous formula that lets you immediately write down answers to problems like this is:

- The number of ordered n-tuples $(a_1, a_2, …, a_n)$ of positive integers with $a_1 + a_2 + … + a_n = m$ is $_{m-1}C_{n-1}$.

A nice way to see why this formula works (and to help you reconstruct it if you forget it) is to picture 10 X's lined up in a row. You can make up a three-numbers-adding-up-to-ten example by placing dividers between two pairs of X's. The number of X's to the left of the first divider is a. The number between the two is b. And the number to the right of both is c. For example, the divider representation of $2 + 3 + 5 = 10$ is:

$$X \ X \ X \ X \ X \ X \ X \ X \ X \ X$$
$$\hspace{1.2cm} \wedge \hspace{1.0cm} \wedge$$

If you think about it for a minute, then it becomes obvious that the number of ways to make three numbers add up to 10 is the number of ways to choose two of the nine gaps to put dividers in. This is $_9C_2$.

Here's one variant on this theme to try for practice. Note that the question asks about *nonnegative* integers, not positive integers.

> *Question: How many ways can one choose nonnegative integers a, b, c, and d with a + b+c+d =6.*

A nice way to get the answer is to realize that if we have four nonnegative integers adding up to six, then by adding one to each number we'll get four positive integers adding up to ten. Hence the answer the same as the answer to "how many ordered quadruples of positive integers add up to 10." This is $_9C_3 = 84$.

One application of this formula is counting paths on a grid, e.g.

> *Question: Consider a city that is laid out on a square grid. In this city, the shortest path from (0,0) to (6,3) is 9 blocks long. How many different 9-block-long paths are there between these two points.*

The first step in doing this problem is pretty obvious. A path will be 9 blocks long if and only if all steps are either to the right or up. But how many ways are there to do this? The tricky second step is to note that we can describe any such path by a giving four nonnegative integers: the number of steps to the right you go before taking your first step up, the number of steps to the right you go in between the first and second steps up, …, and the number of the steps to the right you go after your third step up. Note that these four numbers must add up to 6 if the path ends at (6, 3). Hence, the answer is the same as the answer to the above problem: $_9C_3 = 84$.

A related problem is counting how many ways there are to write an ordered sum when they don't specify how many terms are in the sum:

> *Question: Three distinct ways to write 10 as an ordered sum of two or more positive integers are 5+5, 1+4+5 and 1+5+4. How many such sums are there?*

The way to do this problem is to again think of generating such sums by putting marks in a row of 10 X's.

$$X \ X \ X \ X \ X \ X \ X \ X \ X \ X$$
$$\scriptstyle \land \qquad\qquad \land \ \ \land$$

Now though, you're unrestricted as to how many dividing marks you put in. Counting the number of possible sums remains easy. You have two choices for the gap between the first two X's. Put in a mark or not. Whatever you do here you have two choices for the gap between the second and third X's: put in a mark or not. This makes you think the answer is $2^9 = 512$. This is close, but wrong. The answer is 511 because they said that we must write 10 as a sum of *two or more* positive integers. This means that the expression (10) that comes from putting in zero marks doesn't count.

4.6 Counting Problems Without an Easy Answer: Organize Your List

Some MathCounts hotshots also take the AMC 10. Not infrequently they do quite well on the AMC 10—the AMC 10 is a lot like a hard MathCounts test—and get invited to take the AIME. This can be a humbling experience. AIME problems are so much harder than MathCounts problems that the star students taking it still only average about 4 correct answers *in three hours.* (There are 15 problems on the test.). What MathCounts students aren't used to when taking the AIME is that most problems don't have easy answer. Instead of looking for some simple way to do the problems in your head, you have to accept that there isn't one. Instead, you should look for some way to solve the problem and do it even if it's going take twenty minutes.

In the last few years MathCounts has started to have some problems of this variety. The key is to be organized. Come up with some way to organize your calculations and stick to it. The best way to give a feel for this is with some sample problems. I'll start with one from the 2009 AMC 8:

> *Question: How many noncongruent triangles can be formed using points from the regular 2 x 4 grid below as vertices?*

You can't do this problem by just plugging into some formula you know. You could try to solve it by just starting to draw triangles and seeing how many different ones you can come up with, but this is a risky strategy: it's easy to miss one or to not realize that two triangles that look different are actually the same but just rotated differently.

One way to count these triangles more reliably would be to make an organized list. An organization strategy that often works well with triangles is to organize by the lengths of the sides: this is a good idea because any two triangles with the same side lengths are congruent. In the figure above, there are six possible side-lenghs: $1, 2, 3, \sqrt{2}, \sqrt{5}$, and $\sqrt{10}$. If I were solving the problem, I'd rearrange these from smallest to largest, $1, \sqrt{2}, 2, \sqrt{5}, 3, \sqrt{10}$, and then start checking whether each possible combination of a smallest side and second smallest side could be found. For example, if we want to make the smallest and second smallest sides 1 unit long, then we must have a vertical and horizontal segment. This is possible in a 1-1-$\sqrt{2}$ right triangle, so I put $\sqrt{2}$ in the third column of the first row and put "Yes" in the fourth column. Next, I look for a triangle with smallest sides 1 and $\sqrt{2}$. The 1-1-$\sqrt{2}$ triangle doesn't count, but I can make a 1-$\sqrt{2}$-$\sqrt{5}$ triangle by taking the first two points in the top row and the third point in the bottom row. It turns out that every possible combination starting with a 1 works until you come to $\sqrt{10}$, which isn't possible because you'd need to have a big isosceles triangle. Starting with $\sqrt{2}$ as the smallest side we find three more. Beyond this, there are no more solutions. If you try to make the smallest side 2 units or more, then there's no place to put the third point that isn't closer than two units away from one of the endpoints of the original segment. I'll just put a … on my list instead of writing out all these nonworking combinations. In MathCounts you should do this too – you want to be careful, but you can't afford to be too careful under MathCounts time limits. The answer is 8.

Smallest	2nd Smallest	Largest side	There?
1	1	$\sqrt{2}$	Yes
1	$\sqrt{2}$	$\sqrt{5}$	Yes
1	2	$\sqrt{5}$	Yes
1	$\sqrt{5}$	$\sqrt{10}$	Yes
1	3	$\sqrt{10}$	Yes
1	$\sqrt{10}$	--	No
$\sqrt{2}$	$\sqrt{2}$	2	Yes
$\sqrt{2}$	2	$\sqrt{10}$	Yes
$\sqrt{2}$	$\sqrt{5}$	3	Yes
$\sqrt{2}$	3	--	No
$\sqrt{2}$	$\sqrt{10}$	--	No
2	2	--	No
…			

A few other problems that you want to think about in similar ways are:

Question: In how many ways can 9 be written as a sum of two or more positive odd integers? For example, three such ways are 5 + 3 + 1, 5 + 1 + 3, and 3 + 3+ 3.

Question: How many distinct products can be obtained by multiplying two distinct numbers from the set {-9, -6, -3, 0, 3, 6, 9}?

Question: How many two-element subsets of the set {1, 2, 3, …, 2009} have the property that the sum of their elements is a multiple of 10.

It would probably be useful for you to think to yourself how you'd organize the calculations for these problems before reading my suggestions.

My suggestion for the first one would be to organize a list on the basis of what the largest number used (and the second- and third-largest). For example, if you want to use a 7 the sum must be 7 + 1 + 1 in some order. There are three such orderings. Next, if you want to use a 5 and a 3 the sum must be 5 + 3 + 1 in some order. There are 6 of these. (Look back at section 4.4 "Using combinations to count orderings if you're having trouble keeping track of these.) If you want to use a 5 and no 3's the sum must be 5 + 1 + 1 + 1 + 1. This gives 5 more. Next, move on to sums without 5's. There is just one sum involving 3 3's: 3 + 3 + 3. With two 3's you need to do $3 + 3 + 1 + 1 + 1$ in some order. There are $_5C_2 = 10$ of these. Continuing this process I eventually get a total of 33 ways to do it.

My suggestion for the second one would be to have the first organization be products that are positive, negative, and zero. The zero products are the easiest—zero is the only number equal to zero. Positive products are also pretty easy: you get these by multiplying together two numbers from the set $\{3, 6, 9\}$. (You could use two negatives, but this just gives the same answers.) This gives 3 products. Finally, you get negative products my multiplying one number from $\{3, 6, 9\}$ by one from $\{-3, -6, -9\}$. I'd just make a 3 x 3 table to see that 6 of the 9 products are distinct. The answer is 10.

My suggestion for the third problem would be to focus on the numbers mod 10. First, you can do it by picking numbers that are both multiples of 10. There are $_{200}C_2$ such subsets. Next, you can do it by picking two numbers that are congruent to five mod 10. This gives another $_{201}C_2$ subsets. Finally, you can do it by picking one number congruent to x mod 10 and another congruent to $10-x$ mod 10 for any x in $\{1, 2, 3, 4\}$. This gives 4×201^2 more subsets. In total this adds up to 201604.

In many problems of this type there are multiple ways to go about organizing a list. The key is just to pick some reasonable way and stick with it until you get to the end. I also recommend trying to organize your work very neatly on your scrap paper. If it's a team round it's very important that you be able to show someone else on your team exactly how you did the problem so they can check to make sure you didn't miss any combinations. And even on an individual round I think it makes it less likely that you'll make a mistake. (It also gives you a way to check your work in the unlikely event that you have enough time to go back and do this.)

4.7 Probability Problems Involving Combinations

The probability problems I discussed in connection with IMLEM Meet #5 could usually be done either by thinking about a grid of possible outcomes or by using conditional probabilities. In MathCounts meets they've often asked questions that are most easily done using your knowledge of combinations. For example,

Question (IMLEM Meet #5, Apr. 2001): Two different integers from 1 to 6 are chosen at random with all combinations being equally likely. A fraction is formed by putting the smaller number on top of the larger number. What is the probability that the fraction can be reduced?

Question: Ping pong balls labeled with the numbers 1 to 100 are put in a bag. Three balls are taken out of the bag. What is the probability that the numbers on the three balls are consecutive?

Question: Three distinct grid points from the regular 5 × 5 grid shown below are chosen at random. What is the probability that the three points are collinear?

These problems may seem very hard and very different from problems that are about rolling two dice, but they're not. The only difference is that instead of having an N × N matrix of equally likely events you have a different set of equally likely events. All you need to do to get the answer is to figure out the number of equally likely events, and then figure out how many of them have the desired property.

In the first problem, for example, note first that there are $_6C_2 = 15$ ways to choose the two numbers. The fractions that can be reduced are 2/4, 2/6, 3/6, and 4/6. Hence, the answer is 4/15.

The second problem is similar. There are $_{100}C_3 = \dfrac{100 \times 99 \times 98}{3 \times 2 \times 1}$ ways to choose the three numbers. Each combination is equally likely. Hence, the probability of getting consecutive numbers is the number of combinations in which the numbers are consecutive divided by $_{100}C_3$. The number of combinations involving consecutive numbers is 98: (1,2,3), (2,3,4), (3,4,5) … (98,99,100). The answer is 98 / $_{100}C_3 = (98 \times 6) / (98 \times 99 \times 100) = 1/1650$.

At a conceptual level the third question is no harder. The answer is $P = \dfrac{\# \, of \, collinear \, combinations}{total \, \# \, of \, combinations}$. The denominator is easy. The number of ways to choose three points on the grid is $_{25}C_3 = \dfrac{25 \times 24 \times 23}{3 \times 2 \times 1}$. The problem is harder than the first two, however, because the numerator also involves a tricky counting problem. One good way to do it is to think of how many collinear lines of different lengths there are on the grid. First, there are 12 lines of length five: five horizontal lines, five vertical lines, and two diagonal lines. One way to get three collinear points is to choose any three points on one of these lines. There are $_5C_3 = 10$ ways to choose three points on teach of these lines, so this gives 120 combinations. Second, there are 4 lines of length four (they're diagonals above and below the two main diagonals). There are $_4C_3 = 4$ ways to choose three points on each of these lines, so this gives another 16 combinations. Finally, there are 16 lines that intersect three points on the grid (three each with slopes ½ and -½, two each with slopes 1 and -1, and three each with slopes 2 and -2). Each gives one more combination of three collinear points. The answer is

$$\frac{120 + 16 + 16}{(25 \times 24 \times 23)/(3 \times 2)} = \frac{8 \times 19}{25 \times 4 \times 23} = \frac{38}{575} \, .$$

When you get older you may have a chance to go to ARML. It's the largest (sort of) face-to-face contest in the U.S. A couple elite schools go on their own, but mostly the contest is for regional and state all-star teams. About 2000 participants gather at four test sites for a two day contest. One of the great traditions of ARML was problem 8. Problem 8 was their solution to a difficult problem: "how can you give an 8 question individual test to a group of 2000 students which includes most of the members of the U.S. national math team without ending up with a massive tie among students who get all 8 questions right?" Their solution was to give one very, very hard problem. Specifically, you pick problem 8 so that only about 5 of the 2000 students will be able to solve it in the time allotted. (You get 10 minutes to do each pair of individual problems.) The problem below is a sample:

Question: A random graph on 5 vertices is formed as follows. For each pair of vertices a fair coin is tossed. If the coin comes up heads the two vertices are connected by an edge. If it comes up tails they are not. What is the probability that a random graph formed in this way will be connected? (A graph is connected if there is a path from every vertex to every other vertex.)

I'm not going to tell you the answer. This would ruin the fun. The problem can be done with the material I've discussed in this section and bit of thinking and a lot of inclusion-exclusion counting. I'd be very surprised if anyone reading this book could do it in just 5 minutes though. ARML has recently changed from having 8 questions to having 10 questions. I assume they'll now make question 10 really hard.

4.8 Binomial Probabilities and Pascal's Triangle

Mathcounts not infrequently has problems about multiple coin flips. For example,

Question: A fair coin is tossed five times. What is the probability that Heads comes up exactly twice?

This question is very similar to the problems from the previous section. There are $2^5=32$ equally likely sequences that can occur because you can put an H or a T in each of the five places. The number that involve exactly two heads is $_5C_2 = 10$ – you need to pick which two of the five places to put the H's in. Hence, the answer is $10/32 = 5/16$.

Binomial probability problems are problems like this, but where all outcomes are not equally likely. For example,

Question: A unfair coin is twice as likely to land on Heads as on Tails. The coin is tossed five times. What is the probability that Heads comes up exactly twice?

There's a formula you can memorize to do these problems, but I think it's better to just understand where it comes from. We can compute the probability of getting exactly two heads by adding up the probabilities of all of the sequences that have two heads, e.g adding the probability of getting HHTTT, the probability of getting HTHTT, the probability of HTTHT, and so on. The probability of any single sequence is easy to compute. For example, the probability of getting Heads on the first two tosses and tails on the next three with the unfair coin is $\frac{2}{3} \times \frac{2}{3} \times \frac{1}{3} \times \frac{1}{3} \times \frac{1}{3} = 4/3^5$. The number of such sequences is $_5C_2 = 10$. Hence the answer is $10 \times 4/3^5 = 40/243$.

If you prefer memorizing things to understanding where they come from a formula is:

- If an event happens with independent probability p on each of N trials, then the probability that it occurs on exactly k of the N trials is $_NC_k p^k (1-p)^{N-k}$.

MathCounts sometimes likes to make problems a little harder than this by using conditioning. For example they might ask:

Question: A fair coin is tossed five times. At least one of the tosses results in heads. What is the probability that Heads comes up exactly two times?

The answer to this problem is $_5C_2/31 = 10/31$ instead of $_5C_2/32$ because they told you that one of the 32 possibilities that would have been in the denominator, TTTTT, did not occur.

One problem solving tip for problems like this is that you should keep in mind that Mathcounts problems are designed so that they can be solved by (very clever) kids who do not know things like how to compute binomial probabilities. It could be that this is because the numbers are not so large as to make it impossible to list every possibility. Or it could be that there's some neater trick. Here's one example.

Question: A test has 25 true-false questions worth 4 points each. Bob knows the answer to the first twenty questions and guesses and then randomly on the last five. What is the probability he will score at least 90 on the test?

If you know binomial probabilities you will be tempted to proceed as follows: Getting at least 90 points requires that Bob guesses correctly on at least 3 of the last 5. Hence, you can compute the probability by computing the probability of getting 3 right, the probability of getting 4 right, and the probability of getting 5 right, and adding these up. This method is not so bad if you know all the $_5C_x$ numbers from Pascal's triangle and write them down quickly: $10/32 + 5/32 + 1/32 = 16/32 = \frac{1}{2}$.

The simple answer reflects that there's a better way to do this problem. The problem is asking you to find the probability of 3, 4, or 5 Heads on five tosses. By symmetry, we know that the probability of 5 Heads is the same as the probability of 5 Tails, which is the probability of 0 Heads. Similarly, the probability of 4 Heads is the same as the probability of 1 Head. The probability of 3 Heads is the same as the probability of 2 Heads. Hence, the probability of 3, 4, or 5 Heads is the same as the probability of 0, 1, or 2 Heads. These add up to one, so each must be one-half.

Be a little careful with this trick though. You could get yourself confused about what's symmetric. For example, the probability of getting 4 or more Heads on 8 tosses is not one-half. It's a little more than one-half ($\frac{1}{2} \, _8C_4 /2^8$ more to be precise). Note also that it was important that the coin was a fair coin. If the coin came up Heads with probability two-thirds the probabilities aren't symmetric and you couldn't use a trick like this. How careful you should be depends on where you see a problem like this. On a chapter or state individual round they probably are trying to reward kids who are clever enough to see the symmetry trick. At nationals or on a state team round, on the other hand, they probably figure that all kids will have seen the symmetry trick and may be trying to see which kids will fall for the trap of thinking it's symmetric and which know how to compute binomial probabilities.

Algebra

MathCounts is more algebra-heavy than are IMLEM meets. A number of the questions are designed so you can attack them with or without algebra, but it really, really helps if you're so comfortable with algebra that you naturally use it to solve problems without even thinking about it. As in the IMLEM sections, it doesn't really make sense for me to try to cover a whole year's worth of algebra here. If you don't know algebra well you should go over all the topics covered in a standard middle school book and practice, practice, practice. That will help a lot more than reading this section. In this section discuss a few topics where I have something to say that might not have been covered well in your algebra course. But I'm just going over a few things that come up occasionally, not the standard algebra problems that are much more common.

5.1 Proportional Reasoning

The first thing I wanted to say about algebra is that a number of problems that look like algebra (or geometry) problems are not actually algebra problems: they are proportional reasoning problems. Some sample problems are

Question: Right triangle ABC has legs 5 and x. Triangle DEF has sides 15, 3x, and $\sqrt{225 + 9x^2}$ *. If the area of triangle ABC is 12, find the area of triangle DEF.*

Question: John runs at 6 minutes per mile. Steve runs at 7 minutes per mile. How long will it take Steve to finish a race that takes John 14 minutes.

Question: AB is a diameter of circle C. Suppose that D is a point on the segment AB and AD:DB = 2:1. Let C' be a circle with center D that contains the point B. If the area of circle C is 24, what fraction of the area of C is outside C'?

Question: The wheels on Kate's bike have a 20 inch diameter. The wheels on Anna's bike have an x inch diameter. When Kate and Anna go on a ride together Kate's wheels make 1200 revolutions and Anna's wheels make 1000 revolutions. What is x?

These problems are designed to get you to start mindlessly doing algebra. The area of triangle ABC is 12. This tells me that ½ 5 x = 12, so x=24/5. The Pythagorean theorem tells me that the other triangle is a right triangle. Its legs are 15 and 72/5. So the area is ½ 15 (72/5) = 108.

The solution above is a fine solution, but involves much more work than is really necessary. The main proportional reasoning facts are:

- If shape X is similar to shape Y and the sides (diameter, perimeter, etc.) of shape X are k times as long as the corresponding sides of shape Y then the perimeter of shape X is k times as long as the perimeter of shape Y, the area (resp. surface area) of shape X is k^2 times as large as the area (resp. surface area) of shape Y and the volume of shape X (if it's 3-dimensional) is k^3 times as large as the volume of shape Y.

In the first problem above, the sides of triangle DEF are three times as long as the sides of triangle ABC, so its area will be $3^2 = 9$ times the area of triangle ABC. $9 \times 12 = 108$.

In the second question, you don't need to figure out how many miles long the race is. You just need to know that it takes John 7/6 times as long to run any race. This race takes him $(7/6) \times 14$ = 49/3 minutes = 16 minutes and 20 seconds.

In the third question, AD:DB = 2:1 means that the radius DB of circle C' is one-third of the diameter of circle C, which makes it two-thirds of the radius of circle C. Hence, the area of circle C' is 4/9 of the area of circle C. The answer to the problem is 5/9.

In the final question, the fact that Anna's wheel makes 5/6's as many revolutions means that it's circumference must be 6/5ths the size of Kate's. The diameter is therefore also 6/5 times as large or 24 inches.

MathCounts occasionally uses the phrase "x is inversely proportional to y" in a problem. This means that x is proportional to $1/y$. For example, the 2008-2009 *Handbook* had a problem in which they said that the force that must be applied to a wrench is inversely proportional to its length. More familiar examples are that the number of revolutions a wheel makes is inversely proportional to the radius and that the time needed to travel a given distance is inversely proportional to the speed at which one travels.

5.2 Polynomial Division with Remainders

In fourth grade you learned long division with remainders. They probably didn't say it this way, but mathematically what you learned when they showed you how to compute $d\overline{)n}$ was how to find a quotient q and a remainder r with $0 \le r < d$ such that $n = q \times d + r$. Polynomial division is basically the same idea:

- If p(x) is a polynomial of degree m and d(x) is a polynomial of degree n, then there exist unique polynomials q(x) and r(x) with r(x) having degree less than n such that p(x) = q(x) d(x) + r(x).

For example, if we divide $x^2 + 3x + 3$ by $(x + 1)$ we find that the quotient is $(x+2)$ and the remainder is 1, i.e. $x^2 + 3x + 2 = (x+2)(x + 1) + 1$. The process of doing long division with polynomials is just like the process you learned in 4th grade. For example, in this problem you'd compute the answer as

$$x + 1 \overline{\smash{)}\begin{array}{l} x + 2 \\ x^2 + 3x + 3 \end{array}}$$

$$\underline{x^2 + x}$$
$$2x + 3$$
$$\underline{2x + 2}$$
$$1$$

One place where this comes in handy is in answering questions like

Question: Suppose $x^2 + x = 3$. What is $x^3 - 4x$?

If you don't know about polynomial division his problem may be perplexing. You could use the quadratic formula on the first equation to solve for x. But it's something with a square root in it, so computing x^3 will be a big pain. And are you sure that there is an answer that doesn't depend on which root of the quadratic you choose?

If you think about polynomial division, the problem is no longer daunting. Suppose we find a quotient $q(x)$ and a remainder r(x) so that $x^3 - 4x = q(x) (x^2 + x - 3) + r(x)$. Then, for any x for which $x^2 + x = 3$, it's obvious that $x^3 - 4x$ will be equal to $r(x)$. So all we need to do is some long division

$$x^2 + x - 3 \overline{\smash{)}\begin{array}{l} x - 1 \\ x^3 \quad\quad - 4x \end{array}}$$

$$\underline{x^3 + x^2 - 3x}$$
$$- x^2 - x$$
$$\underline{- x^2 - x + 3}$$
$$- 3$$

The remainder is -3, so the answer to the problem is -3. Thinking about the polynomial division also makes is clear that it was something of a coincidence that the problem had a unique solution. Normally, when you divide a cubic by a quadratic the remainder is a linear polynomial. If the remainder had an x in it, then the answer would depend on which root you chose. Here, we get a unique solution because the remainder turns out to be polynomial of degree zero instead of a first degree polynomial.

5.3 Systems of Equations

In some years MathCounts has lots of problems involving systems of equations. In the IMLEM Meet #2 and Meet #4 Algebra sections I stressed the importance of knowing your category: when the category is equations in one unknown something that looks like a system can be solved one equation at a time; and when the category is linear equations a question that looks like a cubic equation isn't really a cubic equation. MathCounts doesn't have categories, but the same comment applies. You can think of the category as "Big systems of equations that are really just single-variable problems and systems of two linear equations in two unknowns."

Here are a couple sample problems of the first type.

Question: The Gunderson family has three children: Bob, Mary, and Sam. Bob is two years older than Mary. Five years ago, Sam was half as old as Mary. Next year, the sum of the three children's ages will be thirty-five. How old is the oldest child?

Question: Mrs. Gunderson baked some cookies and left them on the kitchen table. Bob came by and ate two fewer than one-third of the cookies. Then Sam came in and took half of the remaining cookies. Mary then took two-thirds of the remaining cookies. Bob came back and noticed that the number of cookies remaining was 10 fewer than he had left on the table. How many cookies did Mrs. Gunderson bake?

In the first problem it's pretty obvious that you should focus on Mary's age. The second and third sentences tell us that $B = M + 2$ and $(S - 5) = \frac{1}{2}(M - 5)$. The latter simplifies to $S = \frac{1}{2}M + 2.5$. The sentence about next year tells us that $(M + 2 + 1) + M + 1 + (\frac{1}{2}M + 2.5 + 1) = 35$. The solution to this is $M=11$. At this point, be sure to double-check what it was that the question asked. It asked for the age of the oldest child. That's Bob, who is 13.

The second problem is a bit harder. What the variables should be and which one to solve for is less obvious. A general tip for questions like this is that it's typically a good idea to start from the end and work backwards. And the variable you often want to work with is the number of things left over. Let T be the number of cookies on the table at the end. The fact that Mary took two-thirds of the cookies tells us that $M=2T$ and there were 3T cookies when she got there. Given that Sam took half of the cookies that were there when he arrived, this means that there were 6T cookies when Sam got there. This is where Bob originally left things, so the last sentence tells us that $6T - T = 10$ or $T=2$. Now, back to the question they asked. Bob ate two fewer than one-third of the cookies leaving 12 cookies on the table. If the answer isn't obvious you can find it by solving $(\frac{1}{3}C - 2) + 12 = C$.

Problems that require you to solve two equations are usually more straightforward. An example might be

Question: Find the intersection (x, y) of the lines $3x + y = 5$ and $x - 2y = 2$. Write your answer as an ordered pair of mixed numbers.

To do this you obviously just need to remember how to solve systems of two equations in two unknowns. The one thing I wanted to point out is that MathCounts problems that have $(1\frac{5}{7}, -\frac{1}{7})$ as an answer will always tell you whether to write your answer using mixed numbers or common fractions. If you get an answer like this to a MathCounts problem that doesn't say whether to use mixed numbers or common fractions, don't think to yourself "Oh. I guess I can write the answer any way I want." Instead you should think "Oh. I guess I got the problem wrong and should redo it. If they don't tell me whether to write my answer as a mixed number or common fraction, the correct answer must only involve integers."

5.4 Maximizing Quadratic Functions

One final thing to remember is that a quadratic equation takes on its maximum (or minimum) value at the point that is the average of the two roots. For example, consider

Question: What is the smallest possible value for y if x is a real number and y = (x − 4)(x − 10)?

Question: For what value of x does −x^2 + 4x + 5 achieve is maximum possible value?

The answers are -9 and 2. The first quadratic is minimized at $x = 7$, where its value is -9. The second can be factored as $-(x + 1)(x − 5)$, so the maximum occurs at $x = 2$.

Another way to do the second one is to "complete the square": $-x^2 + 4x + 5 = -(x − 2)^2 + 9$. This expression is obviously maximized when $x = 2$.

Made in the USA
Monee, IL
28 May 2021